Electronic
Logic Circuits

Electronic Logic Circuits
Third Edition

J. R. Gibson
Department of Electrical Engineering and Electronics
University of Liverpool

Newnes

OXFORD • AUCKLAND • BOSTON • JOHANNESBURG • MELBOURNE • NEW DELHI

Newes
An imprint of Butterworth-Heinemann
Linacre House, Jordan Hill, Oxford OX2 8DP
225 Wildwood Avenue, Woburn, MA 01801-2041
A division of Reed Educational and Professional Publishing Ltd.

A member of the Reed Elsevier plc group

First published in Great Britain 1979
Reprinted with solutions to selected problems 1980
Second edition 1983
Third edition by Arnold, 1992
Reprinted by Butterworth-Heinemann 2002
Transferred to digital printing 2004
© J R Gibson 1992

British Library Cataloguing in Publication Data
Gibson, J. R.
    Electronic logic circuits - 3rd ed.
    I. Title
    621.3815

ISBN 0 340 54377 9

For more information on all Newes publications please visit our
website at www.newespress.com

Typeset by Wearside Tradespools, Boldon, Tyne & Wear

Preface

Large complex electronic logic circuits are now common components in systems ranging from small low cost domestic items to large complex industrial plants. Twenty years ago the design of logic circuits was a specialist subject taught only to students of electronic engineering towards the end of their courses, today some students have used logic circuits before starting formal engineering courses and logic circuit design is met by students of most engineering disciplines. This book is based on courses given to first and second year students in the Department of Electrical Engineering and Electronics at the University of Liverpool. The courses are introductory ones for students who later specialize in digital electronics and related subjects; they also provide background knowledge for those who will only be indirectly involved with logic systems.

The basic features of logic systems and the theoretical concepts on which they are based are constant so the large part of the previous editions concerned with these fundamentals is retained in this new edition. However there are several areas where changes have occurred which affect the way in which logic circuits are implemented. The main causes of change arise from advances in integrated circuit technology; there have been significant changes since the previous edition was prepared. For example, the introduction to the more complicated types bistable through the master-slave structure is retained in this edition as this is a basic concept. However almost all available integrated circuit bistables are now edge-triggered types and usually only devices of this type are available for exercises, therefore the examination of these devices has been extended. A second change is the increased use of ASICs and the availability of reusable user programmable types. It is now possible for students to perform complete exercises from the initial specification to the final circuit testing using these devices. Attempts to create ASIC design exercises before such devices were available often were little more than elaborate computer games.

One other area of change, and one that creates many difficulties, is in the choice of symbols to represent logic circuit elements. Most countries have accepted IEC publication 617, Part 12: 'Binary Logic Symbols' as the basis for their national standards. These standards are complicated and the underlying concepts are beyond the scope of an introductory text; a simplified approach related to that used by manufacturers in their data books and similar publications has been adopted. This is briefly outlined when the symbols are introduced and there are further comments in Appendix C. Anyone who has to prepare formal specifications of logic circuits in a commercial environment must consult the relevant standard for the country in which they are working.

The elementary approach of the previous editions has been retained, in places it is slightly extended, as the subject is met at a very early stage in many engineering courses. Although some students have relevant background knowledge many do not. Several topics found in more advanced courses are omitted or are introduced by a single example of an application to illustrate basic ideas. For example logic functions are introduced directly, rather than through set theory and Venn diagrams. The direct approach to the

Boolean functions AND, OR and NOT appears to cause no difficulties when it follows an explanation of the engineering advantages of two-stage components. A simple approach is also used for sequential circuit design and this is described in terms of a single lengthy but reliable technique.

Methods for the design of logic circuits form a significant proportion of this book and, as any design technique can only be completely learnt by practice in its use, problems are included at the end of most chapters. It is also essential that students should use standard components to construct and test circuits they have designed. Those who attend courses forming the basis for this book undertake at least ten hours related laboratory work, most undertake significantly more.

I wish to thank my colleagues, students, former students and many others for discussions and questions with which they unknowingly assisted in the preparation of this book. Thanks are also due to Motorola Semiconductors Inc., for their help. Finally the assistance of my wife Kathleen, and of Mary Ballin and Joan Palmer in the preparation of the original edition is appreciated.

J.R. Gibson
1991

Contents

Preface		**v**
1	**Two-state systems**	**1**
1.1	Two-state systems	1
1.2	Electronic two-state systems	2
1.3	Common electronic logic components	3
1.4	Applications	5
1.5	Numbers	5
1.6	Codes and coding	6
1.7	Problems	8
2	**Basic elements of combinational logic**	**10**
2.1	Truth tables	10
2.2	A set of basic logic elements	12
2.3	Boolean arithmetic	18
2.4	Boolean algebra	20
2.5	de Morgan's theorem	23
2.6	Duality	23
2.7	The exclusive-OR function	24
2.8	Logic circuit analysis	26
2.9	Problems	27
3	**The design of combinational logic circuits**	**29**
3.1	Development of a truth table	29
3.2	Minterms and maxterms	31
3.3	Minterm representation of circuits	32
3.4	Minimization	33
3.5	Karnaugh maps	34
3.6	Minimization using Karnaugh maps	36
3.7	Circuit implementation	39
3.8	Unspecified states	41
3.9	Summary of the design method	43
3.10	Particular maps	44
3.11	Multiple-output circuits	44
3.12	Comment	46
3.13	Problems	46
4	**Sequential logic elements**	**48**
4.1	The set–reset flip-flop	48
4.2	The operation and use of an SR flip-flop	50
4.3	The clocked SR flip-flop	51

4.4	The D-type flip-flop (or latch)	53
4.5	The serial shift register	54
4.6	Other registers	56
4.7	The JK flip-flop	57
4.8	Edge-triggered bistables	60
4.9	Operation of an edge-triggered bistable	61
4.10	Comments	64
4.11	Problems	64
5	**Sequential logic systems**	**66**
5.1	Counters	66
5.2	Pulse sequences	68
5.3	Asynchronous pure binary counters	69
5.4	An application of an asynchronous counter	71
5.5	Elimination of transient outputs	72
5.6	Problems	73
6	**The design of sequential logic circuits**	**74**
6.1	Some definitions	74
6.2	State diagrams	75
6.3	State tables	77
6.4	Development of an excitation table	78
6.5	Design of the circuit	81
6.6	Summary of the design method	83
6.7	Problems associated with unused states	87
6.8	Design from timing diagrams	90
6.9	Comments	92
6.10	Problems	92
7	**Electronic logic circuits**	**94**
7.1	Timing problems	94
7.2	Interconnections	97
7.3	Special interconnection techniques	100
7.4	External connections	102
7.5	General assembly points	104
7.6	Problems	105
8	**Large logic networks**	**107**
8.1	Addition of binary numbers	107
8.2	The cell technique for combinational circuit design	111
8.3	Use of cells for large sequential circuits	112
8.4	Other large sequential circuits	114
8.5	LSI devices	117
8.6	Comments	118
8.7	Problems	119
9	**Application specific integrated circuits (ASICs)**	**120**
9.1	Approaches to ASIC manufacture	120

9.2   User programmed ASICs   121
9.3   Combinational circuits   122
9.4   Sequential circuits   125
9.5   Concluding remarks   128

Appendix A   Number systems and base conversions   **129**

Appendix B   Maxterm representation of circuits   **133**

Appendix C   A note concerning symbols   **136**

Appendix D   Solutions to selected problems   **138**

Bibliography   **144**

Index   **145**

1 Two-state systems

If some condition of a manufactured article may change or be changed then the article may be classified either as a discrete-state or as a continuous-state system. The term 'system' simply describes the article as a whole; the article may be a component used in some large assembly or it may be a complete large assembly. The 'state' of the system is its condition, each different state corresponds to a different condition of the system. A system is a continuous-state system when its variable condition, its state, is able to take any value between certain limits; such systems have an infinite number of possible states. For instance the sound output from a radio receiver may be adjusted by the listener to any level from inaudible to the maximum possible output. Similarly, a car driver can select any engine speed required by use of the accelerator control. Both the radio receiver and the car engine are examples of continuous-state systems; such systems are also known as continuously-variable systems.

In contrast to the wide range of states (conditions) possible in a continuous-state system, a conventional electric light switch allows only on or off settings of the light; similarly, a car driver can use the gear lever to select one of the small number of gear ratios available. Both the lighting system with an on–off switch and the car gearbox are examples of discrete-state systems; in such systems only a finite, usually fixed, number of different states are allowed.

Many situations arise in which a designer has to choose to use either a discrete-state or a continuous-state system because it is possible to solve a particular engineering problem by using either type of system. For instance an on–off light switch may be replaced by a dimmer control which allows the user to select any light level over a wide range. Because designers frequently have to choose to use either a discrete-state or a continuous-state system they should be aware of the particular advantages and disadvantages of each type of system.

A general comparison for every possible situation cannot be made, but in a large proportion of cases the continuous-state system is more expensive to manufacture and is less reliable than a discrete-state alternative. For example the cost of a dimmer light control is typically five times that of an on–off switch. However, the dimmer control has the advantage that the user may select any level of lighting rather than just fully on or fully off. This wider range of choice given to the user is found in many, but not all, cases when a continuous-state system is chosen rather than a discrete-state system.

1.1 Two-state systems

Discrete-state systems which have only two possible states, for example a simple on–off light switch, are particularly easy to manufacture in many cases. This book is concerned only with systems which are constructed entirely from discrete-state components which are restricted to two possible states, i.e. the components are themselves **two-state systems**. Surprisingly, this very strict limitation placed on the components imposes only one restriction on the complete system; it must be a discrete-state system, but it may have any number of states because two-state components may be combined together so that they form a multiple-state system.

Two-state components are often called **logic elements** and complete systems which are constructed from such components are **logic systems**, **logic circuits** or **logic networks**; the reasons for these names should become apparent in Chapter 2. An on–off light switch is a two-state device as are many other types of switch; an alternative name for logic circuits is **switching circuits**. One further name is **digital circuits**.

Nearly all the systems (circuits) which will be described in this book can be constructed using any two-state device as the basic component but the use of electronic two-state components is emphasized. Systems built using electronic logic components have become familiar everyday items; before the early 1970s such systems were only used in specialized engineering applications. The application of electronic devices to a wide range of new and existing products followed the development of modern electronic logic elements which have many advantages when compared with other types of logic or switching component. They are exceedingly small, require very little power, operate quickly, and are extremely cheap. The cost of a single logic element within a system containing many elements is often less than 0·001 pence. Manufacturing techniques and component design appear to be continuing to advance rapidly; it is probable that element costs will continue to decrease, typically halving every three to five years.

An additional advantage of modern electronic components is that those used in two-state systems have exceptionally high reliability. Well-constructed systems containing several thousand million components will operate with over 10^9 element switchings per second for periods of several years without any component failures. If the components were not electronic (e.g. mechanical springs and levers, hydraulic devices or electromechanical relays) failures would occur so often in such large systems that they would rarely, if ever, operate successfully.

1.2   Electronic two-state systems

There are many ways to construct electronic circuits so that they are two-state circuits. The most common circuit designs are such that any circuit input or output which is above some specified voltage level is defined to be in one of the two states. When an input or output is below some much lower level it is in the other state. It is essential that there is a well-defined gap between the two voltage levels. All the logic circuits (and those connected to them) must be designed so that all the inputs and outputs can only take up the specified voltage levels; they must never be held in the region between the two levels. Figure 1.1 illustrates this method of defining the logic states, the voltage levels shown are those most commonly used at the present time. Note the difference indicated in the levels for inputs and outputs. The requirement that an output '1' voltage must be higher than the value recognized as '1' at an input, and that an output '0' is lower than the value of an input '0' ensures that when the output of one circuit is connected so that it supplies the input to another circuit the level will always be correctly recognized.

Most real circuits have construction features which restrict how much higher than the minimum 1 level, or how much lower than the maximum 0 level, input voltage levels may be without damage occurring. Although inputs are not allowed to be set to values between the two levels a change of an input from one level to the other requires that the voltage passes through the unallowed (transition) region. Components are designed to

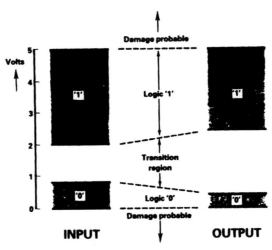

Fig. 1.1   Definitions of logic states (TTL levels)

ensure that no problems occur provided that the change through the transition region is made quickly and smoothly.

It is convenient to give names to the two logic states so that they are easily identified. If one works entirely with such names then any logic network designed may be constructed using logic components of any single type. The nomenclature adopted in Fig. 1.1 is that the upper level (state) is called 1 (ONE) and the lower level is called 0 (ZERO). Alternative pairs of names in common use include HIGH and LOW, UP and DOWN, ON and OFF, and TRUE and FALSE. The names 1 and 0 are those which will be used in this book, although others will be used if they are particularly suited to some application.

Detailed consideration of the many ways in which electronic circuits may be designed to behave as two-state elements are to be found in books concerned with the design and analysis of electronic circuits. Any reader interested in such details should refer to one of the many books on these subjects. This book is concerned with the design and behaviour of circuits which are constructed from components, where each component is a complete two-state system which is normally – but not necessarily – an electronic circuit.

1.3   Common electronic logic components

With the exception of a few special applications, modern electronic logic networks are constructed from two-state components which are puchased in the form of integrated circuits. A logic integrated circuit is an electronic circuit consisting of components (resistors, transistors, etc.) connected so that the resulting circuit functions as the required logic device. The complete circuit is fabricated on the surface of a small piece of a single crystal of a very pure semiconductor material (the piece of crystal is often called a chip). Most devices are made with silicon as the semiconductor material and the chip is typically 1 to 10 mm square and about 0·5 mm thick. The chip is encapsulated in a protective casing of plastic or ceramic which is larger than the chip. Metal leads are fixed to the chip and extend beyond the plastic or ceramic case to allow electrical connections to be made to the circuit inputs, outputs and power supply points. This packaging arrangement enables the integrated circuit to be easily handled and connected into a larger network by hand or by automatic machines. Figure 1.2 is an outline drawing of one of the most common encapsulations for small and medium size integrated circuits; the dual in line (DIL) package.

At present the relatively simple electronic logic components required for initial studies of logic circuits are readily available in two distinctly different circuit forms; transistor–transistor logic (TTL) and complementary symmetry metal oxide semiconductor logic (CMOS). The two systems are different and components of the different systems should not be interconnected unless the complete circuit specifications are understood. However, when this is necessary, special interconnecting (interface) circuits are used to make connections between the two different systems. As adequate ranges of components are available in each of the common systems, devices from a single system can be used for most designs and the interconnection problem may be avoided.

The TTL and CMOS systems supersede several much earlier ones, for example resistor–transistor logic (RTL) and diode–transistor logic (DTL). There is such a large investment in the manufacture of TTL and CMOS components, and in equipment that uses them, that these types will probably be the ones most commonly used for some time. However, even these types are regularly modified; devices with revised (improved) specifications become available at regular intervals. For applications requiring very high speeds of operation (above 100 million operations per second) components using emitter-coupled logic (ECL) are required. Great care and a knowledge of high speed circuit layout techniques are necessary to build circuits to operate at such high speed.

When the various component types were first introduced it was possible to describe their differences easily. Early TTL components were much faster than early CMOS ones but required much more power. However, TTL devices have been developed so that modern versions use much less power than earlier forms, also modern CMOS devices now operate much more quickly than older forms. Further, there are many different families of both types available and each family has been designed to have characteristics making it suitable for certain applications. Except in Chapter 7, the various types will not be considered and no assumptions will be made concerning the type of two-state components required for circuit construction. In nearly all cases any type of component – including non-electronic ones – may be used.

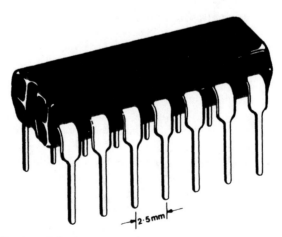

Fig. 1.2   DIL packaged integrated circuit

Most students of logic system design will find that the construction and testing of logic circuits is an essential part of their studies. For this purpose one of the '7400' families of components will be the most useful as they are easily obtained at low cost. Most, but not all, are also difficult to damage by incorrect handling or connection. In addition to a small selection of components, some form of prototyping system is required; there are several readily available at a reasonable cost.

1.4 Applications

Two-state (logic) circuits are used in many devices and systems. The field in which they were first widely used is that of computers, calculators and related machines and it is with these devices that they are most commonly associated. However, their use extends to a wide range of control and test equipment; applications include telephone-exchange equipment, railway signalling, aircraft flight control and lift (elevator) controls. Complex electronic logic systems are now used in a wide range of consumer products; these include washing machines, audio and video equipment, automobile engine management systems, automobile braking controls and many others.

The dominant position of number processing machines (e.g. computers and calculators) as the major initial application of logic circuits, has greatly influenced the development of components and the methods used to describe logic circuits. Consequently some knowledge of a few basic features of number processing machines is necessary. In particular some understanding of number systems and of coding is required when developing logic circuit design techniques.

1.5 Numbers

To understand how a system constructed entirely from two-state components may be used to manipulate numbers it is necessary to examine the principles of number systems. Most people are trained when they are children to use numbers and to perform arithmetic operations but, quite reasonably, very few people are taught the basic theory of numbers and arithmetic.

It is customary to use the decimal number system and to anyone familiar with the system the printed form of a number such as $2093 \cdot 54$ has an exact and easily understood meaning. This printed form is a shorthand representation of

$$2 \times 1000 + 0 \times 100 + 9 \times 10 + 3 \times 1 + 5 \times 0 \cdot 1 + 4 \times 0 \cdot 01$$

which can also be written as

$$2 \times 10^3 + 0 \times 10^2 + 9 \times 10^1 + 3 \times 10^0 + 5 \times 10^{-1} + 4 \times 10^{-2}$$

This form is just one example of a much more general form of number which is

$$d_n \times b^n + d_{n-1} \times b^{n-1} + \cdots$$
$$+ d_2 \times b^2 + d_1 \times b^1 + d_0 \times b^0 + d_{-1} \times b^{-1} + d_{-2} \times b^{-2} + \cdots$$
$$+ d_{-m} \times b^{-m}$$

where b is the **base** (or radix) of the number system and the ds are the **digits**. In any number system which has this form, the number of different digits required is equal to the base of the system, i.e. the number of different symbols for digits is equal to the base.

The origin of the decimal system which has a base of ten is obvious; simple counting and arithmetic is often performed using fingers (digits) to keep a record; each finger therefore represents a different symbol in the number system and a base of ten is a natural consequence.

An advantage of this type of number system is that only a few different symbols (different digits) are required and this small number of symbols may be used to represent any number no matter how large or how small it is. This is possible because the position of any digit relative to the decimal point is important; a digit l places to the left of the point is multiplied by the base to the power $l-1$, i.e. by b^{l-1} and a digit r places to the right is muliplied by the base to the power $-r$, i.e. by b^{-r}. The digits themselves, usually 0, 1, 2, 3, etc., up to a value one less than the base, are just printed symbols; they provide one particular method of indicating the quantities zero, one, two and so on. There is no reason why the symbols must be these printed ones; provided that a clearly defined and consistent method is developed any type of system could be used to represent a digit. One possible scheme is to use the different conditions (states) of a discrete-state system to represent the digits.

When the digits of some number system are represented by the states of a discrete-state system, each state will be used to represent a different digit and the system must have as many states as there are digits in the number system. Therefore, when a two-state system is used to represent digits there can only be two different digits and the number system must have a base of two. The digits in such a system are usually called 0 (zero) and 1 (one) as they must represent nothing and a single unit respectively. These digits may be represented by the two states of a logic element and the reason for calling the states of such an element 0 and 1 is now obvious. The number system which has a base of two is called the **binary system** and most electronic computing devices operate in the binary system or in a modified form of it. Some details of the binary system and frequently used other bases, including simple methods of converting numbers in one base to their equivalent in another, are given in Appendix A.

Printed binary numbers seem strange when they are first encountered, but once understood they are easier to manipulate than decimal numbers. The most obvious feature of binary numbers – other than the fact that they consist entirely of ones and zeros – is the large number of digits that are required to represent quite small quantities; for example the binary number 101101110 is equivalent to the decimal number 366. Whenever a binary number is used and it is not obvious that it is binary rather than decimal, the base will be indicated by a subscript; for example 101_2 is the binary number with the value five.

1.6   Codes and coding

Although the binary system is the one which is usually used to represent numbers associated with logic systems, it is not essential to use this number system. Logic elements may be combined to form multiple-state systems and the states of such systems may be used to represent numbers in systems other than the binary one. When multiple-state systems are built from two-state components the state of each component may be considered to represent one **binary digit** which is usually called a **bit**. If all the bits are considered simultaneously as a binary number with several digits, then each different number represents one state of the multiple-state system. For example when a system includes three bits denoted by A, B and C there are eight possible different states; some of these are given in Table 1.1

Table 1.1

C	B	A
0	0	0
0	1	0
1	1	0
1	0	1
1	1	1

The bits may be considered to be a code in which each combination (i.e. each row in Table 1.1) represents one of the states of a multiple-state system. If the bits of some multiple-state system are required to represent the ten decimal digits then a large number of different codes are possible. A convenient code is one in which four bits A, B, C and D are treated as the digits of a binary number with A as the least significant (2^0) digit and D as the most significant (2^3) digit. Table 1.2 is a listing of this code.

Table 1.2

Bit (Weight)	D ($2^3 = 8$)	C ($2^2 = 4$)	B ($2^1 = 2$)	A ($2^0 = 1$)	Decimal digit
	0	0	0	0	0
	0	0	0	1	1
	0	0	1	0	2
	0	0	1	1	3
	0	1	0	0	4
	0	1	0	1	5
	0	1	1	0	6
	0	1	1	1	7
	1	0	0	0	8
	1	0	0	1	9

Each row in Table 1.2 is the binary code for a single decimal digit; because each bit represents a definite power of two the full name of this scheme is 'the eight-four-two-one weighted binary coded decimal representation'. This is often shortened to just 'binary coded decimal' or 'BCD'. Although many other codes such as 5421 and 2421 weighted binary codes are used to represent decimal numbers, the use of the initials BCD without any qualification implies the 8421 weighted code.

Codes may be constructed in many ways and often have no form of weighting. For unweighted codes there is no specific value for each bit and the code must be interpreted by a look-up approach. Table 1.3 is one example of such codes for the six digits 0 to 5.

The code in Table 1.3 is a Gray code and is one of many Gray codes which can be constructed for six digits. These codes have the property that the number of bits which have the value 1 always increases or decreases by one, and only one bit changes, when a move is made from one position to the next in the code sequence. The code in Table 1.3 has several symmetrical features and is a special case known as a reflected Gray code. (A different Gray code without the symmetry of the reflected form is obtained if the code value for digit 2 is replaced by 101.)

Gray codes have advantages over other codes in applications when the code values are read in sequence by mechanical or optical reading devices. These codes are used extensively on the code plates of position measuring systems in automatic equipment, particularly automatic machine tools.

Table 1.3

C	B	A	Digit
0	0	0	0
0	0	1	1
0	1	1	2
1	1	1	3
1	1	0	4
1	0	0	5

Coding is a convenient method of indicating the condition or state of a multiple-state system constructed from two-state components. Such coding is frequently used to describe logic circuits but the representation is only occasionally referred to as a code. It is assumed by authors of component data books, technical literature and similar material that their readers will understand that a binary code is being used.

1.7 Problems

1 Which of the following contain at least one component which is a discrete-state system?
   a) A door lock.
   b) A water tap (control valve).
   c) A pendulum clock.
   d) A thermostat (temperature control) for an oven or a water heater.
   e) A controller for traffic signals at a road junction.
   f) The braking system of a bicycle or car.

2 Read Appendix A then perform the following conversions of numbers in one base to their equivalent in the alternative base stated.
   a) Decimal numbers 357, 68, 59·72 and 23·375 to binary.
   b) Binary numbers 11010, 1011, 1110111·101 and 101010·011 to decimal.
   c) Octal number 276 to decimal.
   d) Binary number 11101110 to octal.
   e) Hexadecimal numbers 3CA and F4·7D to binary.
   f) Decimal numbers 1021 and 59·72 to hexadecimal.

3 Devise a four-bit binary code to represent the ten decimal digits. The code must be such that if it is interpreted as a binary number this number would have a value three greater than the decimal digit represented by the code. (This is called the excess three code.)

4 What is the minimum number of binary digits (bits) required to represent a single octal (base eight) digit? Devise a code for all the duodecimal (base twelve) digits using the smallest number of bits necessary.

2 Basic elements of combinational logic

Systems constructed so that all the inputs and outputs can only take either one of two allowed states were introduced in Chapter 1 and termed logic systems or circuits. If each output of such a system depends only on the present states of the inputs to the circuit it is called a **combinational logic circuit**. In a combinational system there is no dependence of one output on the other outputs. Also input states which have occurred previously have no influence on the circuit behaviour. An alternative way of stating this last fact is that the order in which the inputs are applied to the circuit does not affect its final output. Such a system is shown schematically in Fig. 2.1.

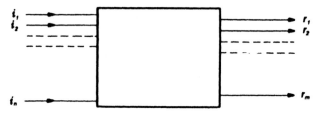

Fig. 2.1   Combinational logic network

In the general case illustrated by Fig. 2.1 there are n inputs $i_1, i_2, i_3, \ldots, i_n$ and m outputs or results $r_1, r_2, r_3, \ldots, r_m$; each input may take either one of the two logic states and the outputs are also restricted to these logic levels. The actual level at a particular output at any time depends on the logic states present at all of the inputs at that instant in time.

It is assumed that as soon as any input changes then the outputs change immediately to the levels which correspond to the new input conditions. Any real circuit will take a finite time to operate; this time is called the **propagation delay** and it may be neglected in most simple applications of combinational logic circuits.

2.1   Truth tables

Detailed examination of the general case of a system with a large number of outputs is not necessary. It is sufficient to consider the case of a system with several inputs and a single output, R. The general case of a circuit with m outputs is just m separate single output circuits which all have the same inputs connected to them; i.e. all the circuits have the same n inputs but each has a different output. Figure 2.2 illustrates this showing a 4 input and 3 output circuit; each box represents a separate, and different, combinational logic circuit with four inputs and single output.

When a circuit has n inputs 2^n different input situations may arise because each input may be 0 or 1 and all possible combinations of input states may exist. For example if

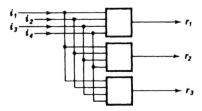

Fig. 2.2 Multiple output circuit built from single output circuit modules

there are three inputs A, B and C, then there are eight different input situations; all eight are shown in Table 2.1. In this table there is a separate column for every input and each row corresponds to one of the possible combinations of input states.

Table 2.1

C	B	A
0	0	0
0	0	1
0	1	0
0	1	1
1	0	0
1	0	1
1	1	0
1	1	1

For each input situation the output, R, must take the value 0 or 1. An n input circuit has 2^n possible combinations of input states and for each of these R must be either 0 or 1; hence 2^{2^n} different combinational logic circuits exist which have n inputs and a single output. Table 2.2 shows all sixteen possible circuits which have only two inputs; each of the columns headed R_1, R_2, . . . , etc., corresponds to the output of a different circuit.

Table 2.2

Inputs		Possible Outputs														
A	B	R_1	R_2	R_3	R_4	R_5	R_6	R_7	R_8	R_9	R_{10}	R_{11}	R_{12}	R_{13}	R_{14}	R_{15}	R_{16}
0	0	0	0	0	0	0	0	0	0	1	1	1	1	1	1	1	1
0	1	0	0	0	0	1	1	1	1	0	0	0	0	1	1	1	1
1	0	0	0	1	1	0	0	1	1	0	0	1	1	0	0	1	1
1	1	0	1	0	1	0	1	0	1	0	1	0	1	0	1	0	1

Obviously a particular circuit with two inputs and a single output behaves so that its output corresponds to one of the output columns of Table 2.2. The behaviour of any single-output combinational logic circuit may be described by a table which is similar to Table 2.2 but has only one output column. Table 2.3 is an example of one possible table for a circuit with three inputs.

Table 2.3 is called a **truth table** and it completely specifies the behaviour of a particular logic circuit. The rules for constructing a truth table are that there must be separate columns for each input and for each output (as indicated previously, circuits may have more than one output but can be divided into separate single-output circuits).

The table must have one row for each possible combination of input states, every possible input combination must be included and the output(s) produced by the circuit must be shown in all cases.

A truth table is a clear and easily understood method of describing the behaviour of a combinational logic circuit. Since it also provides an exact definition of the circuit operation, a truth table should be included as part of the specification of any combinational logic circuit.

Table 2.3

Inputs			Output
C	B	A	R
0	0	0	0
0	0	1	1
0	1	0	1
0	1	1	0
1	0	0	1
1	0	1	0
1	1	0	0
1	1	1	0

2.2 A set of basic logic elements

It was stated earlier that 2^{2^n} different logic circuits may be specified which have n inputs and a single output. Each one of these circuits has its own unique truth table. If n is 2 there are sixteen possible circuits, when n is 3 there are two hundred and fifty-six and so on. The number increases rapidly with n and it would be an uneconomic proposition to manufacture different circuits for every possible case. Fortunately it is possible to choose a set consisting of a small number of basic logic elements and then interconnect several of them to produce a circuit which behaves in the manner required by some specification.

There is no unique set of basic logic elements but the five which follow are probably the most useful ones; in addition they directly implement a set of mathematical rules which may be used to describe the behaviour of logic systems.

The basic logic elements used in combinational logic systems are usually called 'gates' by electronic engineers. This is because one common application is to control the flow of electrical logic signals along some path; the element may be used to allow the signal to pass or may prevent its passage and this action is analogous to opening or closing a gate across a path.

2.2.1 The AND gate

This element has a *single output and any number of inputs*. It is defined as the combinational logic circuit which gives the output 0 unless all the inputs are at the logic 1 level; in this case only the output is 1. This definition is sufficiently precise to allow the truth table of an AND gate to be written down immediately. Table 2.4 is the truth table for a three-input AND gate.

Table 2.4

Inputs			Output
C	B	A	R
0	0	0	0
0	0	1	0
0	1	0	0
0	1	1	0
1	0	0	0
1	0	1	0
1	1	0	0
1	1	1	1

When this element is used as a component in a large system it is convenient to be able to represent it by some quickly recognized symbol; this makes interpretation of circuit diagrams of a network drawn using such symbols relatively simple. Figure 2.3 shows several different symbols which are used to represent an AND gate. In all cases the symbols have been drawn with the inputs from the left (three inputs are shown) and the output towards the right; this is the recommended practice that should be adopted when drawing circuit diagrams.

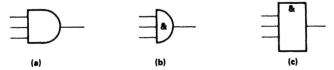

(a)   (b)   (c)

Fig. 2.3   AND gate symbols

The rectangular symbol of Fig. 2.3c is now the accepted standard for a complete electrical circuit which behaves as an AND gate (it may be combined with further symbols to indicate additional electrical features). Previously the symbol of Fig. 2.3a was used to represent an electrical AND gate circuit element; however modern practice is to use this symbol to represent the logical AND process for all methods of implementing the logic function. Another way of stating this is that the rectangular symbol represents an electrical circuit while the symbol represents a 'pure logical' AND function.

The practice that will be adopted here follows that of current integrated circuit data books and computer based circuit design systems. The rectangular symbol is used to represent a complete electronic AND gate circuit. The dee symbol is used to indicate the AND function when describing a network constructed of logic components when these may be of any type. This has the obvious disadvantage that students new to the subject must become familiar with two sets of symbols; also the difference in their use is rather subtle. However it is necessary to be familiar with both as data sheets and instruction manuals follow this practice. Nearly all the circuits (networks) met before Chapter 7 are considered in a general form, consequently the dee symbol is the one most frequently used. Diagrams of networks using symbols indicating only logic (e.g. dee for AND) will be referred to as logic circuit diagrams. When networks require a specific electrical component implementing a logic function the rectangular symbol will be used and the network will be referred to as an electrical circuit. Further brief comments concerning symbols are made in Appendix C.

One further method which is used to describe any logic circuit is to express its behaviour in terms of Boolean algebra. This algebra is the mathematical description of systems which are only allowed to have two states and it will be examined in more detail later. An essential definition in Boolean algebra is the AND function, and in equations the function is represented by a dot. For example, the Boolean expression for a three-input AND gate is the equation

$$R = A.B.C$$

which is read as R *equals* A *and* B *and* C.

2.2.2   The OR gate

This element is another one which implements a fundamental definition in Boolean algebra. The OR gate is the combinational logic circuit which has any number of inputs and has an output of 1 when any one, or more than one, of the inputs is 1. An alternative statement is that the circuit gives the output 1 unless all the inputs are 0, in which case it gives the output 0. Note that the output of a two-input OR circuit is not just 1 in the cases input A is 1 with B equals 0, or input B is 1 with A equals 0; it also includes the situation where A and B are both 1 (i.e. A is 1 or B is 1 or A and B are both 1). From this description the truth table can be constructed; Table 2.5 is that for a three-input OR gate.

Table 2.5

Inputs			Output
C	B	A	R
0	0	0	0
0	0	1	1
0	1	0	1
0	1	1	1
1	0	0	1
1	0	1	1
1	1	0	1
1	1	1	1

Figure 2.4 shows some of the symbols used to represent an OR gate. Following the practice introduced for AND gate symbols the form in Fig. 2.4c represents a complete electrical OR gate unit while that in Fig. 2.4a is the symbol for a 'pure logical' OR function. Note that if a circuit diagram uses one form of symbol for AND it must only use the corresponding OR symbol and vice versa. That is, the two forms of symbol should

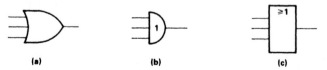

(a)   (b)   (c)

Fig. 2.4   OR gate symbols

not be mixed; a diagram is entirely a logic circuit or it is entirely an electrical circuit.

In Boolean algebra the OR function is represented by a plus sign and the equation describing a three-input example is

$$R = A + B + C$$

which is read as R *equals* A *or* B *or* C.

2.2.3   The INVERTER or NOT gate

Unlike the two gates described previously this gate is only defined in the case in which it has a single input and a single output whereas the others may have any number of inputs. This is the one gate which is essential in any basic set of logic functions. Its action is to produce an output logic state which is not the same as the state at the input. Because only two states are possible in a logic system this statement is a complete description. The truth table of an inverter is obvious and is shown in Table 2.6.

Table 2.6

Input A	Output R
0	1
1	0

Many different symbols are used to represent this gate and several shown in Fig. 2.5. Again the rectangular symbol is used for a complete electrical unit while the triangular one is used in diagrams which only describe the logic behaviour of a circuit. Symbols for this gate may be constructed in several other ways as inversion is indicated by a small circle on logic circuit diagrams. On electrical circuit diagrams inversion is indicated by a small triangle, as in Fig. 2.5d. The two methods of indicating inversion have different meanings, a general discussion of the difference between them is beyond the level of this book. However, if the small circle is only used on logic diagrams and the small triangle is only used on electrical diagrams the results will generally be correct.

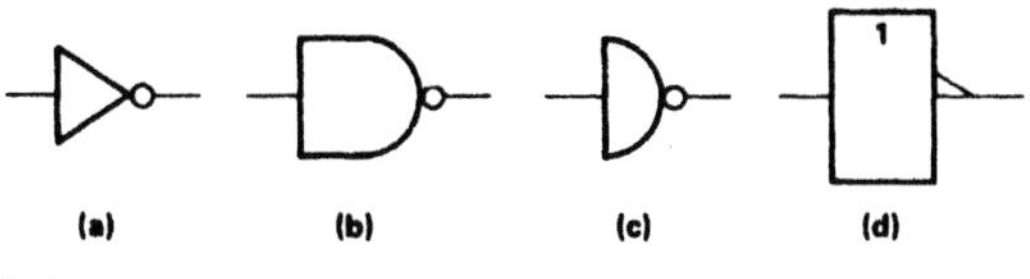

Fig. 2.5 INVERTER symbols

The method of representing inversion in Boolean algebra is to draw a line above the quantity to be inverted. Therefore, if R is the result of inverting A the equation which describes this operation is

$$R = \bar{A}$$

which is read as R *is not* A, or alternatively as R is the *inverse* of A. The inverse of a logical quantity is also known as its **complement**; in the equation above R is the complement of A.

2.2.4   The NAND gate

This gate is equivalent to an AND gate followed by an inverter (NOT gate). In Boolean algebra it is conventional to state the final NOT before the quantity which is inverted. Therefore this gate is said to perform the operation NOT–AND which is usually contracted to NAND. While the derivation of the name can cause confusion regarding the action of the gate the fundamental AND then INVERT definition requires that the output of a NAND gate must be exactly opposite to the output of an AND gate. Therefore a NAND gate will always have an output of 1 except when all the inputs are 1; if all the inputs are 1 then the output is 0. Table 2.7 is the truth table for a two-input NAND gate.

Table 2.7

Inputs		Output
B	A	R
0	0	1
0	1	1
1	0	1
1	1	0

This gate may be represented by the symbol for an AND gate followed by that for an inverter as in Fig. 2.6a. However, as NAND gates are widely used the contracted forms shown in other parts of Fig. 2.6 are normally used. The selection of a small circle or of a small triangle to indicate inversion was briefly discussed when inversion was introduced in Section 2.2.3.

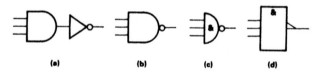

(a)   (b)   (c)   (d)

Fig. 2.6   NAND gate symbols

In terms of Boolean algebra this is written as

$$R = \overline{A.B}$$

which is R equals not the result of A and B, or alternatively, R *equals NAND of* A *with* B. The statement 'R equals not A and B' is not clear – it could mean that R is the quantity $\overline{A}.B$ or that it is the quantity $\overline{A.B}$.

When a line is drawn above a quantity the inversion must not be performed until the quantity beneath the line has been evaluated; i.e. the quantity beneath the line should be considered to be in brackets and should be completely evaluated before the inversion operation is performed.

Example 2.1

The quantity $R = \overline{A.B.C}$ is $R = (\overline{A.B.C})$ and this is not the same as $X = \overline{A}.\overline{B}.\overline{C}$ which is $X = (\overline{A}).(\overline{B}).(\overline{C})$. Prove that the quantities X and R are different.

Solution

Table 2.8

| Inputs | | | Intermediates | | | | Outputs | |
C	B	A	$\bar{C}$	$\bar{B}$	$\bar{A}$	$Y = A.B.C$	$R = \bar{Y}$	$X = (\bar{A}).(\bar{B}).(\bar{C})$
0	0	0	1	1	1	0	1	1
0	0	1	1	1	0	0	1	0
0	1	0	1	0	1	0	1	0
0	1	1	1	0	0	0	1	0
1	0	0	0	1	1	0	1	0
1	0	1	0	1	0	0	1	0
1	1	0	0	0	1	0	1	0
1	1	1	0	0	0	1	0	0

The proof consists of preparing a table then giving reasons, based on the table, why the logical relationship proposed must be true. The table consists of a column for every initial (input) quantity, a column for intermediate quantities formed after every logic operation, and a column for each of the results (output quantities). There is a separate row in the table for every possible combination of values of the input quantities involved.

For the example the intermediates are the inverse values $\bar{A}$, $\bar{B}$ and $\bar{C}$ as well as the quantity $Y = A.B.C$; the outputs are R and X which are to be proven to differ. Table 2.8 is the result; for every value of A the column for $\bar{A}$ is completed from the definition for the Boolean NOT function. Similarly $\bar{B}$ is obtained from B, and $\bar{C}$ from C. Using the definition of the AND function the column for Y is completed, R is obtained from Y using the definition of NOT. Finally X is obtained from the values of $\bar{A}$, $\bar{B}$ and $\bar{C}$ using the definition of AND. The output columns show R and X for **every possible** set of conditions of A, B and C that can occur. These output columns were obtained using only the definitions of the Boolean operations and as the two columns are different then $R \neq X$.

To some degree this proof is too extensive as a logic relationship must be untrue if any single case is found which is untrue. However the same technique may be used to prove two logical quantities to be identical, for such a proof the relationship must be true for every possible case. This method of proving logical relationships by examination for every combination of input values is sometimes referred to as perfect induction.

2.2.5 The NOR gate

This gate is equivalent to an OR gate followed by an inverter; i.e. the NOR function

Table 2.9

| Inputs | | Output |
B	A	R
0	0	1
0	1	0
1	0	0
1	1	0

is NOT–OR. The NOR gate has an output of 0 unless all the inputs are 0, in which case the output is 1. The behaviour of a two-input NOR gate is given in Table 2.9. In Boolean algebra this is written as $R = \overline{A + B}$ which is R equals not the result of A or B, or alternatively, R *equals NOR of A with* B.

The OR symbol can be modified to show inversion, just as the AND symbol was, by a small circle or a small triangle after the symbol for OR to produce the symbol for a NOR gate. Several NOR gate symbols are shown in Fig. 2.7. As for NOT and NAND the small circle and small triangle do not have identical meanings.

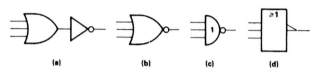

(a)   (b)   (c)   (d)

Fig. 2.7   NOR gate symbols

As in the case of the NAND function, care is required in stating the inversion operation; for example, $\overline{A + B + C}$ is not the same as $\overline{A} + \overline{B} + \overline{C}$ and the difference can be shown by writing out complete truth tables for both expressions.

Note that the truth tables for a single-input NAND gate and a single-input NOR gate are identical and are the same as the truth table for an inverter. It is quite common for designers of large logic systems to connect all the inputs of a multiple-input NAND or NOR gate together (it then becomes a single-input gate) to provide an inverter. This construction is common when it simplifies the layout of a circuit. It is usually used because integrated circuits are often supplied with two or more multiple-input gates in the same package; if some of the gates in a package are not required in a particular application it may be economical to use them as inverters.

2.3   Boolean arithmetic

It has already been stated that logic circuits may be described mathematically using Boolean algebra; this originates from work by the nineteenth century mathematician George Boole. The form of algebra which should already be familiar to readers is based on conventional arithmetic involving numbers which may take any value from minus infinity to plus infinity. Boolean algebra is similar but it is based on the arithmetic of logic values.

In Boolean arithmetic numeric quantities may only have either of the two logic values true and false; these have already been given alternative names of 1 and 0. Instead of the four conventional arithmetical operations of add, subtract, multiply and divide the three Boolean operations of AND, OR and NOT are used. These operations have already been defined and are summarized in Table 2.10. Equivalence or identity symbols are used to indicate that terms on each side of a relationship are identical. For example $1 + 0 \equiv 1$ indicates that $1 + 0$ may be replaced by 1 as the result is identical; however it is also the case that 1 may be replaced by $1 + 0$ (1 may also be replaced in many other ways). Each of the relationships in Table 2.10 may be shown to be correct by applying the definition of a particular Boolean operation.

Table 2.10   Boolean arithmetic operations

$$\left.\begin{array}{r}\overline{1}\equiv 0\\ \overline{0}\equiv 1\end{array}\right\} \text{NOT}$$

$$\left.\begin{array}{r}0.0\equiv 0\\ 0.1\equiv 1.0\equiv 0\\ 1.1\equiv 1\end{array}\right\} \text{AND (two inputs)}$$

$$\left.\begin{array}{r}0+0\equiv 0\\ 0+1\equiv 1+0\equiv 1\\ 1+1\equiv 1\end{array}\right\} \text{OR (two inputs)}$$

The relations in Table 2.10 indicate that the order of evaluating terms in a Boolean operation is not important, as indicated $0+1\equiv 1+0$ and $0.1\equiv 1.0$. Relationships involving a single Boolean operation may be extended to any number of terms; either directly from the definition of the Boolean operations, or by using identities in Table 2.10. For example using $1\equiv 1+0$ to replace the left hand 1 of $1+1\equiv 1$ by its equivalent $1+0$ produces $1+0+1\equiv 1$. Extension to other cases is trivial and may be used to demonstrate that the order of terms in a Boolean operation has no effect; results such as the following are easily obtained

$$0+1+1+1\equiv 1+0+1+1\equiv 1+1+0+1\equiv 1+1+1+0$$

$$0.1.1.1\equiv 1.0.1.1\equiv 1.1.0.1\equiv 1.1.1.0$$

These may also be derived by logical argument based on the fundamental definitions of the operators AND and OR.

In more complex Boolean expressions, that is those that involve more than one of the operators AND, OR and NOT, it is necessary to consider the order of evaluation. This problem arises in conventional arithmetic, for example consider

$$3\times 5+4\div 7-2\times 2\times 6$$

In an expression of this form evaluation is performed by working methodically from one end of the expression to the other using each arithmetical operator as it occurs. The EXCEPTION is that when a number has a multiply or divide sign to one side of it and has an add or subtract sign to the other side then the multiplication or division is always performed before the addition or subtraction. That is the example above is more clearly written as

$$(3\times 5)+(4\div 7)-(2\times 2\times 6)$$

where the brackets indicate that quantities inside them should be fully evaluated before the operators outside the brackets are used. It is not necessary to consider the order of evaluation of terms containing only add and subtract, or containing only multiply and divide, as the order of evaluation does not affect the results.

The calculation can be performed without the brackets because the arithmetic operators have a defined order in which they must be used, this order is called **precedence**. For conventional arithmetic the multiply and divide operators have equal precedence; the add and subtract operators also have equal precedence. However multiply and divide have an order of precedence which is higher than that for add and subtract. An order of precedence is necessary for the Boolean operators to allow complicated Boolean expressions to be constructed and evaluated. NOT has the highest order of precedence, that is it is essential to perform NOT operations before any other operations. AND has a lower precedence than NOT but it has a higher precedence than OR. Thus the three Boolean operators have an order of precedence with NOT highest, then AND, and with OR lowest. Although NOT has the highest precedence some care is needed when an overline is used to indicate NOT; everything below the line is inverted in one operation so it is essential to regard everything below an overline as being in brackets to force evaluation before the NOT operation, e.g.

$$1.0.1 + \overline{1.1.0} + \overline{0+1}$$

is

$$1.0.1 + \overline{(1.1.0)} + \overline{(0+1)}$$

Brackets may be used in Boolean arithmetic, exactly as in conventional arithmetic, to override the defined order of precedence. For example the two expressions

$$0.(0+1).(1+0)$$

$$0.0 + 1.1 + 0$$

are not the same although the digits and operators occur in exactly the same sequence in both. The first may be changed to $0.(1).(1)$ by replacing $0+1$ by 1 and $1+0$ by 1 as the terms in brackets must be evaluated first. This then becomes $0.1.1$ which has the value 0 from the definition of the AND function. In the second expression the order of precedence rules require that the two AND terms are evaluated first giving $0+1+0$ which has the value 1. As in conventional arithmetic, brackets may be used within other brackets to any required extent.

2.4 Boolean algebra

Conventional (standard) arithmetic operating on numbers is extended using undefined quantities (variables) to give conventional algebra. In a similar manner Boolean operation on logic quantities is extended to produce Boolean algebra. That is relationships may be written in Boolean algebra containing unknown quantities indicated by some symbol (e.g. letters A, B, C, . . .) where the symbol stands in place of a logic value; that is the symbol represents an unknown quantity which may only be logic 1 or logic 0. As in conventional algebra the quantities represented by symbols may be called variables, they are also known as literals, and to engineers are inputs or outputs when applied to circuits. Table 2.11 lists some examples of relationships which are true in Boolean algebra; equivalence symbols are used as expressions on either side are identical.

Many Boolean relationships appear to be exactly the same as those of similar appearance in conventional algebra, e.g. $A.0 \equiv 0$ and $A+B \equiv B+A$, but others are unusual when first encountered; e.g. $A.A \equiv A$ and $A.(A+B) \equiv A$. It is essential that the Boolean relationships are recognized as differing from conventional algebraic ones and that only the rules of Boolean algebra are used. Although most of the examples in

Table 2.11 have a maximum of three variables the rules may be applied to any number of variables. For instance

$$A.B + A.C + A.D.E \equiv A.(B + C + D.E)$$

and
$$A.\bar{B}.\bar{C}.D \equiv \bar{C}.A.D.\bar{B} \equiv D.\bar{B}.A.\bar{C}$$

All the relationships may be demonstrated to be correct by the method of perfect induction used in Example 2.1. This can be applied to Boolean algebra for any relationship with a fixed number of unknown quantities (variables) as every possible case can be examined. Such a treatment is impossible in conventional algebra as each unknown has an infinite number of possible values.

Table 2.11 Some relationships in Boolean algebra

$A.1 \equiv A$ $A.0 \equiv 0$ $A + 1 \equiv 1$ $A + 0 \equiv A$	Zero and unit rules (Note the dominance of 0 in AND and 1 in OR)
$A.\bar{A} \equiv 0$ $A + \bar{A} \equiv 1$ $\bar{\bar{A}} \equiv A$	Complement relations
$A + A \equiv A$ $A.A \equiv A$	Idempotence
$A + B \equiv B + A$ $A.B \equiv B.A$	Commutative laws
$A + A.B \equiv A$ $A.(A + B) \equiv A$ $A + \bar{A}.B \equiv A + B$	Absorption rules
$A.(B + C) \equiv A.B + A.C$ $A + B.C \equiv (A + B).(A + C)$	Distributive laws
$A + B + C \equiv (A + B) + C \equiv A + (B + C)$ $A.B.C \equiv A.(B.C) \equiv (A.B).C$	Associative laws (insertion of brackets)
$\overline{A + B + C + D + \cdots} \equiv \bar{A}.\bar{B}.\bar{C}.\bar{D}. \cdots$ $\overline{A.B.C.D.\cdots} \equiv \bar{A} + \bar{B} + \bar{C} + \bar{D} + \cdots$	de Morgan's Theorem

Example 2.2

Verify de Morgan's theorem for three variables in the form

$$\overline{A.B.C} = \bar{A} + \bar{B} + \bar{C}$$

Solution

Table 2.12

C B A	$W = A.B.C$	$\overline{W} = \overline{A.B.C}$	$Z = \bar{C}$	$Y = \bar{B}$	$X = \bar{A}$	$X + Y + Z = \bar{A} + \bar{B} + \bar{C}$
0 0 0	0	1	1	1	1	1
0 0 1	0	1	1	1	0	1
0 1 0	0	1	1	0	1	1
0 1 1	0	1	1	0	0	1
1 0 0	0	1	0	1	1	1
1 0 1	0	1	0	1	0	1
1 1 0	0	1	0	0	1	1
1 1 1	1	0	0	0	0	0

Using the method of perfect induction Table 2.12 is prepared. As there are three inputs there are eight possible cases to consider and the table has eight rows. From the definitions of AND and NOT the values of the intermediate quantities $W = A.B.C$, $X = \bar{A}$, $Y = \bar{B}$ and $Z = \bar{C}$ are derived for all possible combinations of input values. Following this further quantities of $\overline{W}$ and $X + Y + Z$ are evaluated using the definitions of NOT and OR; the results are the columns for $W = \overline{A.B.C}$ and $X + Y + Z = \bar{A} + \bar{B} + \bar{C}$. As the two columns are the same for every possible set of values of A, B and C, the quantities $\overline{A.B.C}$ and $\bar{A} + \bar{B} + \bar{C}$ must be identical. This demonstrates that de Morgan's theorem is true in this particular case.

The relationships of Boolean algebra may be used to manipulate Boolean expressions (that is mathematical descriptions of logic circuits) into alternative forms. The usual reason for performing such manipulations is to change one Boolean expression for a circuit into another which is more easily or more economically constructed from the available components.

Example 2.3

Suppose the output, R, of some circuit is given as

$$R = A.C.D + A.\bar{B}.\bar{C}.D + A.B.\bar{C}.D + \bar{A}.\bar{B}.\bar{C}.\bar{D}$$

Can this be manipulated into a more simple form?

Solution

Change the expression using the Distributive laws to

$$R = A.C.D + A.\bar{C}.D.(\bar{B} + B) + \bar{A}.\bar{B}.\bar{C}.\bar{D}$$

Since $B + \bar{B} = 1$ from the Complement relations, then it can be written as

$$R = A.C.D + A.\bar{C}.D + \bar{A}.\bar{B}.\bar{C}.\bar{D}$$

which can be further treated in the same way giving

$$R = A.D.(C + \bar{C}) + \bar{A}.\bar{B}.\bar{C}.\bar{D} = A.D + \bar{A}.\bar{B}.\bar{C}.\bar{D}$$

This final form is much simpler than the original, but some steps in the reduction were not obvious unless the answer was already known.

One fault of Boolean algebra is that the steps in the manipulation are not always obvious. Chapter 3 introduces some methods which may be used to reduce complicated Boolean expressions in a reliable and consistent manner.

2.5   De Morgan's theorem

De Morgan's theorem (or law) is very important; it is probably one of the most frequently used relationships in Boolean algebra. The two forms are

$$\overline{A+B+C+\cdots} \equiv \bar{A}.\bar{B}.\bar{C}.\cdots$$

$$\overline{A.B.C.\cdots} \equiv \bar{A}+\bar{B}+\bar{C}\cdots$$

and both forms are valid with any number of variables.

The principle use of de Morgan's theorem is to convert an OR type of expression (i.e. either OR or NOR) into an AND form (i.e. either AND or NAND) and *vice versa* when a particular type of logic gate is to be used for circuit construction. For example, suppose that $X = A + B + C$ and that a circuit is required to produce X using only NAND gates. Because double inversion is the same as no inversion, then $X = \overline{\overline{A+B+C}}$; applying de Morgan's theorem to $\overline{A+B+C}$ the result becomes

$$X = \overline{\bar{A}.\bar{B}.\bar{C}}$$

Remembering that an inverter is a single input NAND gate then the logic circuit shown in Fig. 2.8 will behave as an OR gate; this circuit is formed entirely from NAND gates.

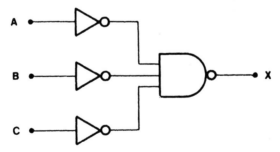

Fig. 2.8   OR function generated by NAND gates

De Morgan's theorem is a mathematical expression which describes the property of all two-state systems which is known as duality.

2.6   Duality

This is a rather difficult concept to grasp and, although it is not essential in the design and use of logic circuits, it is a useful concept in the understanding of basic principles. The treatment here is brief and the reader should consult more advanced texts for more detailed information.

Basically the property of duality is that provided all input *and* output logic levels are inverted, AND becomes OR and OR changes to AND. For example, Table 2.13a is the truth table for a two-input AND gate while Table 2.13b is produced from it by changing every 1 to 0 and every 0 to 1 in both input **and** output columns.

Examination of Table 2.13b shows that it is the truth table of an OR gate, so the process of inverting all logic levels has changed an AND gate into an OR gate. But initially it was necessary to select which logic state was called 1 and which was 0; the choice was entirely arbitrary and there is no reason why the opposite definition cannot be made and used. Hence when the logic levels are defined in one way a particular circuit may behave as an AND gate; when they are defined in the opposite way *the same circuit is an OR gate.*

Table 2.13

Inputs		Output
B	A	R
0	0	0
0	1	0
1	0	0
1	1	1

(a)

Inputs		Output
$\bar{B}$	$\bar{A}$	$\bar{R}$
1	1	1
1	0	1
0	1	1
0	0	0

(b)

A more general form of statement of duality is that if a relationship in Boolean algebra is known to be true then the dual of the relationship is also true. The dual of any quantity or relation can be formed by inverting every variable, changing every AND operator to an OR operator, and changing every OR operator to an AND operator.

For example suppose some quantity X is known to be a function of A, B and C such that

$$X = \overline{(A + \bar{B})}.(\bar{B} + C)$$

The dual relationship may be written directly and must also be true, that is

$$\bar{X} = \overline{(\bar{A}.\bar{B})} + (\bar{B}.\bar{C})$$

2.7 The exclusive-OR function

A basic set of logic elements has been selected; in fact the set is too large because it is possible to construct every possible logic circuit using only NAND gates or only NOR gates (provided an inverter is regarded as a single input NAND gate or a single input NOR gate). A larger range of elements is retained because circuit designs may be more economical when a range is available.

One circuit is required so often that it is convenient to regard it as another logic element. When the OR function was introduced it was emphasized that a two-input OR gate has an output of 1 if either input is at 1 or if both are at 1. An alternative function would be one which gives an output of 1 when either input is 1 but *not* when both inputs are 1. Such a circuit is called an **exclusive-OR** gate (because it excludes the case that both inputs are 1) and unlike the normal OR gate it is only defined as a circuit with two inputs. The truth table for the **exclusive-OR** gate is shown as Table 2.14.

Table 2.14

Inputs		Output
B	A	R
0	0	0
0	1	1
1	0	1
1	1	0

The exclusive-OR function is often indicated by the symbol $\oplus$ in equations. Hence,

$$R = A \oplus B$$

Although the exclusive-OR function is only defined for two inputs, an equation such as $X = A \oplus B \oplus C \oplus D$ is exact because the same result is obtained whichever order the exclusive-OR operations are performed in. That is $(A \oplus B) \oplus (C \oplus D)$ gives exactly the same result as $((A \oplus B) \oplus C) \oplus D$. These multiple-input systems are not called exclusive-OR gates; this name is only applied to the two-input gate.

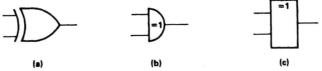

(a)   (b)   (c)

Fig. 2.9 EXCLUSIVE–OR gate symbols

Some of the circuit symbols for exclusive-OR gates are shown in Fig. 2.9. The exclusive-OR function is available as an integrated circuit but it is also easily constructed from the standard functions. For example, it can be shown that using NAND gates

$$R = A \oplus B = \overline{(A.\bar{B}).(\bar{A}.B)} = \overline{(\overline{(A.\bar{B}).A}).(\overline{(\bar{A}.B).B})}$$

and these NAND gate logic circuits are shown in Fig. 2.10.

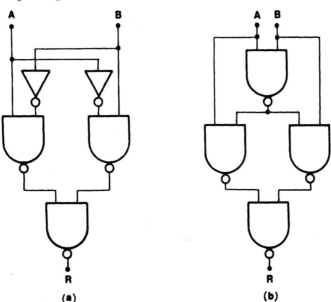

(a)   (b)

Fig. 2.10 NAND gate implementations of the EXCLUSIVE–OR function

2.8   Logic circuit analysis

Those readers who are familiar with linear electronic circuit analysis will be aware that the analysis process is difficult. In the case of combinational logic circuits, analysis is a trivial (although tedious) task. The aim of any circuit analysis is to start from a circuit diagram and to obtain an exact description of the behaviour of the circuit; when the circuit is a combinational logic one the result of an analysis will be either a Boolean expression or a truth table.

To perform the analysis it is sufficient to work systematically through the circuit from the inputs to the outputs, determining the output of every gate.

Example 2.4

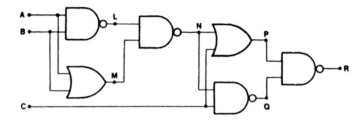

Fig. 2.11

Analyse the circuit behaviour of the combinational logic system shown in Fig. 2.11.

Solution

All the gate outputs have been labelled; it is just necessary to write down a truth table step by step in the way that is indicated by Table 2.15.

Table 2.15

C	B	A	$L = \overline{A.B}$	$M = A+B$	$N = \overline{L.M}$	$P = N+C$	$Q = \overline{N.C}$	$R = \overline{P.Q}$
0	0	0	1	0	1	1	1	0
0	0	1	1	1	0	0	1	1
0	1	0	1	1	0	0	1	1
0	1	1	0	1	1	1	1	0
1	0	0	1	0	1	1	0	1
1	0	1	1	1	0	1	1	0
1	1	0	1	1	0	1	1	0
1	1	1	0	1	1	1	0	1

This technique is tedious but it will always give the correct result. An alternative method is to work in terms of Boolean algebra, i.e. from the circuit diagram,

$$L = \overline{A.B} \text{ and } M = A+B$$

Also,
$$N = \overline{L.M}$$

Hence $N = \overline{L.M} = \overline{(\overline{A.B}).(A+B)} = \overline{\overline{A.B}} + \overline{(A+B)} = A.B + \overline{A}.\overline{B}$

Further $Q = \overline{N.C}$ and $P = N + C$ so that since $R = \overline{Q.P}$ the same manipulations that were used with $N = \overline{L.M}$ give

$$R = N.C + \overline{N}.\overline{C}$$

Hence, $N.C = (A.B + \overline{A}.\overline{B}).C = A.B.C + \overline{A}.\overline{B}.C$

Also, $\overline{N}.\overline{C} = \overline{(A.B + \overline{A}.\overline{B})}.\overline{C} = A.\overline{B}.\overline{C} + \overline{A}.B.\overline{C}$

Therefore the circuit behaviour is described by

$$R = A.B.C + \overline{A}.\overline{B}.C + A.\overline{B}.\overline{C} + \overline{A}.B.\overline{C}$$

The Boolean solution is not always obvious and it is easy to make errors in the manipulation of Boolean expressions. Chapter 3 is concerned with the design (synthesis) of combinational logic circuits and a method by which truth tables may be converted to Boolean expressions is developed. The most reliable circuit analysis consists of producing a truth table for the circuit then converting the table to a Boolean expression by some standard technique.

2.9 Problems

1 Evaluate the following Boolean expressions
   a) $1 + 0.1.0 + 1.1.0$
   b) $0 + 1.(0 + 1 + 0).(1 + \overline{0})$
   c) $(1 + 1.0).(0 + \overline{1}.1).(1.\overline{0} + 1.1)$
   d) $((1.\overline{0} + 1).(1.\overline{1} + 0.(\overline{1 + 0})) + (\overline{1 + 0}).(1.\overline{0}))$

2 Use tabular techniques (perfect induction) to prove that the following relationships are correct:
   a) $(A + B).(A + C) = A + (B.C)$
   b) $A.(A + B) = A$
   c) $A + \overline{A} = 1$
   d) $\overline{A + B + C} = \overline{A}.\overline{B}.\overline{C}$
   e) any relationships in Table 2.11 which do not appear to be reasonable ones.

3 Simplify the following Boolean expressions:
   a) $A.\overline{B}.\overline{C} + A.B.\overline{C} + \overline{A}.\overline{C}$
   b) $M.\overline{N}.P + \overline{L}.M.P + \overline{L}.M.\overline{N} + \overline{L}.M.N.\overline{P} + \overline{L}.\overline{N}.\overline{P}$
   c) $A.B.\overline{C}.\overline{D} + \overline{A}.B.\overline{D} + \overline{A}.\overline{B}.C + \overline{B}.D$

4 Show that the expression $R = \overline{(A.\overline{B}).(\overline{A}.B)}$ is equivalent to the exclusive-OR function.

5 Prove that the following expressions involving the exclusive-OR function are correct:

   a) $A \oplus B = \overline{A} \oplus \overline{B}$

   b) $\overline{A \oplus B} = \overline{A} \oplus B = A \oplus \overline{B}$
   Note that expression (b) is sometimes called an exclusive-NOR function.

6 Derive the truth table and a Boolean expression for the logic circuit shown in Fig. 2.12.

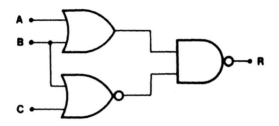

Fig. 2.12

3 The design of combinational logic circuits

The process of design usually begins with a requirement for the development of a product to meet a particular need. This requirement is normally expressed in a specification for the product. Often the specification received by the designer of a combinational logic circuit may be a vague written or verbal statement of the task to be performed by the circuit. Alternatively the specification may be a complete truth table for the circuit, or it may be somewhere between these extremes. Before any attempt can be made to design the circuit a complete specification of the circuit is necessary. To be complete it must describe the exact behaviour of the circuit under all possible circumstances; also the specification must not contain any ambiguities. For a combinational logic circuit the best form of specification is a complete truth table.

3.1 Development of a truth table

The first stage in any systematic approach to the design of a combinational logic circuit is the production of the truth table for the circuit. However, since the initial specification may be given in many different ways, no general rules can be formulated for the construction of the truth table. In common with many design situations much of the process must be left to the experience and common sense of the designer.

To produce a truth table the number of circuit inputs and outputs must be determined in those cases for which they are not explicitly specified. A table of all the possible combinations of input states can then be constructed. Table 2.1 is one example for a circuit with three inputs. This table becomes the truth table when the state of each output is added to the table alongside every input combination. In some cases the specification will impose constraints which influence the form of the truth table to a great extent; in other cases the designer will have a large element of choice in forming the table.

Example 3.1

Devise a truth table for a circuit with the following behaviour. Three simple two-position switches are connected to supply inputs to a logic circuit which has a single output. The output controls a lamp and the circuit operation is such that changing the position of any switch changes the condition of the lamp (i.e. if the lamp was on it goes off and *vice versa*.)

Solution

The specification given is not complete; also the assumption is made that it is possible to design a circuit to perform as required, and this assumption may not be correct. The number of circuit inputs and outputs is defined by the specification but many other features are omitted and must be selected by the designer. This selection produces a revised specification which will allow a complete truth table to be devised for the circuit.

a)   Inputs. The original specification does not state which switch position corresponds to the logic state 0, and which position corresponds to state 1. Arbitrarily choose one position, call it down, and let it represent an input of 0; the other position, up, then represents an input of 1. Also distinguish the three inputs by giving them the labels A, B and C.

b)   Output. Again a free choice of logic states exists, it would seem sensible to choose an output of 1 to cause the lamp to be on and an output of 0 to correspond to the lamp off.

c)   Starting position. Although the specification states how the circuit must change when any input changes it does not indicate the output for any input condition. In this example it is necessary to choose the output for a single set of inputs. A reasonable choice is to decide that if all the inputs are 0 then the output is also 0.

Once these decisions have been made for this particular example the complete truth table can be constructed. Starting with all three inputs at 0 the output is also 0. The specification requires that a change at a single input will change the output. Hence the three different cases with one input of 1 and the other two inputs 0 must all be cases in which the circuit will give an output of 1. A similar argument shows that the three cases with two inputs of 1 and the third 0 must be cases for which the output is to be 0. Finally when all three inputs are 1 the output must be 1 because this is a single switch changed from all of the two 1s and one 0 cases. The complete truth table, Table 3.1, can now be formed.

Table 3.1

Inputs			
C	B	A	Output
0	0	0	0
0	0	1	1
0	1	0	1
0	1	1	0
1	0	0	1
1	0	1	0
1	1	0	0
1	1	1	1

Example 3.2

Obtain the truth table for the circuit which has three inputs A, B and C and one output R. The circuit behaviour is given by the Boolean relationship

$$R = A.(B + \bar{B}.\bar{C}) + \bar{A}.\bar{B}.C$$

Solution

This is another possible form of circuit specification and it is one which leaves very few decisions to be made by the designer. The numbers of inputs and outputs are determined by the specification and the truth table is obtained by evaluating the Boolean expression for every possible set of input conditions. This evaluation is made simple when intermediate quantities (as in Example 2.1) are defined; choose these to be $L = \overline{B}.\overline{C}$, $M = B + L = B + \overline{B}.\overline{C}$, $N = A.M = A.(B + \overline{B}.\overline{C})$ and $P = \overline{A}.\overline{B}.C$. When R is evaluated for all combinations of A, B and C, Table 3.2 is produced.

Table 3.2

Inputs			Intermediates				Output
C	B	A	$L = \overline{B}.\overline{C}$	$M = B + L$	$N = A.M$	$P = \overline{A}.\overline{B}.C$	$R = N + P$
0	0	0	1	1	0	0	0
0	0	1	1	1	1	0	1
0	1	0	0	1	0	0	0
0	1	1	0	1	1	0	1
1	0	0	0	0	0	1	1
1	0	1	0	0	0	0	0
1	1	0	0	1	0	0	0
1	1	1	0	1	1	0	1

3.2   Minterms and maxterms

It is not possible to devise a single formal process which can be used to develop a truth table from all forms of circuit specification, but when the truth table has been prepared it is possible to devise systematic methods of circuit design. These methods require the truth table as the starting-point and a single method is introduced in this chapter; it is one of the most commonly used design techniques.

The first stage in the circuit design procedure is to derive a Boolean expression which describes the required circuit behaviour. This may appear to be unnecessary if, as in Example 3.2, the original specification is itself a Boolean expression for the circuit. However the form of expression in any specification will probably contain an arbitrary mixture of AND, OR and NOT (inversion) operations. To approach the design of combinational logic circuits in a systematic manner a specific form of Boolean expression is required. Therefore the truth table is not just to be converted into a Boolean expression but is to be expressed in one with a particular form.

Because logic functions have a dual nature (see Chapter 2) every technique may be formulated in two different ways. Consequently when a method is devised to produce a logic expression from a truth table there will be a related dual method. Both methods will lead to the same final circuit but one method is usually slightly easier to apply. The method described here is based on minterms; the alternative dual method uses maxterms.

A **minterm** is that AND function which includes the algebraic symbols for every input to the circuit (usually called input variables, variables or literals) once, and only once, in either true or complemented (inverted) form. Further, each minterm corresponds to a single row in the truth table in such a way that the minterm can only have the value 1 when the circuit inputs have the values corresponding to those for that particular row of the truth table. That is each minterm is uniquely associated with a single row in the truth table and may be used to describe that row. Table 3.3 shows all the possible combinations of inputs for a system which has three inputs of A, B and C, together with the minterms which correspond to each case.

Table 3.3

Inputs			
C	B	A	Minterm
0	0	0	$\bar{A}.\bar{B}.\bar{C}$
0	0	1	$A.\bar{B}.\bar{C}$
0	1	0	$\bar{A}.B.\bar{C}$
0	1	1	$A.B.\bar{C}$
1	0	0	$\bar{A}.\bar{B}.C$
1	0	1	$A.\bar{B}.C$
1	1	0	$\bar{A}.B.C$
1	1	1	$A.B.C$

The alternative, or dual, expression is called a **maxterm** and is an OR function which includes each input variable once only in either true or complemented form. Unfortunately there are two different definitions of a maxterm in common use, this causes confusion and is one of the reasons why minterms are used here. The possible forms of maxterms and their use to describe circuits are briefly discussed in Appendix B; the results obtained using maxterms can be shown to be the same as those from minterms.

The extension of minterms (and maxterms) to four or more variables is simple. For example a four-input system will have minterms such as $A.\bar{B}.C.\bar{D}$. Because they have a printed form which is similar to a conventional algebraic product, minterms are often called **products** or product terms. Similarly maxterms may be referred to as **sums** or sum terms.

3.3 Minterm representation of circuits

The conversion of a truth table into a Boolean expression which consists entirely of minterms is straightforward. Each minterm uniquely represents a single row of the truth table. It can only have the value 1 when all the inputs have the values which correspond to those in the row of the truth table associated with it. If all the minterms which correspond to those conditions for which the output must be 1 are listed, and *only* these minterms are listed, then whenever input conditions exist for which the circuit is required to give an output of 1, a single one of the listed minterms will have the value 1. In those cases for which the circuit output is to be 0 all the listed minterms will be 0. The OR function is defined such that it produces the result 1 when any single input (or more than one input) has the value 1. Therefore, if all the listed minterms are combined by an OR function, the complete expression produced is one which exactly describes the behaviour of the circuit represented by the truth table.

Writing the truth table for Example 3.2 (Table 3.2) again without the intermediate columns, but with a column of minterms, Table 3.4 is obtained. The minterms with an asterisk are those which correspond to cases for which the final circuit must produce an output of 1.

Table 3.4

| Inputs | | | Output | |
C	B	A	R	Minterm
0	0	0	0	$\bar{A}.\bar{B}.\bar{C}$
0	0	1	1	$A.\bar{B}.\bar{C}*$
0	1	0	0	$\bar{A}.B.\bar{C}$
0	1	1	1	$A.B.\bar{C}*$
1	0	0	1	$\bar{A}.\bar{B}.C*$
1	0	1	0	$A.\bar{B}.C$
1	1	0	0	$\bar{A}.B.C$
1	1	1	1	$A.B.C*$

In this example, the output must have a value of 1 if any one of the minterms $A.\bar{B}.\bar{C}$, $A.B.\bar{C}$, $\bar{A}.\bar{B}.C$ and $A.B.C$ is 1; therefore the output, R, is given by

$$R = A.\bar{B}.\bar{C} + A.B.\bar{C} + \bar{A}.\bar{B}.C + A.B.C$$

The similarity between this form of expression and one in conventional algebra is such that this is often called a **sum of products**. After the truth table has been developed for a combinational logic circuit, the next step in the systematic design of the circuit is to produce this minterm sum of products expression.

3.4 Minimization

A circuit could be constructed immediately from the sum of products expression by using inverters, AND gates, and one multiple-input OR gate. Such a circuit would operate correctly but would usually be more expensive to manufacture than one in which the Boolean expression had been simplified in some way. In the example above the minterm formulation gave the relationship

$$R = A.\bar{B}.\bar{C} + A.B.\bar{C} + \bar{A}.\bar{B}.C + A.B.C$$

Using the rules of Boolean algebra this may be reduced to

$$R = A.B + A.\bar{C} + \bar{A}.\bar{B}.C$$

This second form obviously requires fewer logic elements. Further, most of the elements required for this reduced form have fewer inputs than those in the original. It is reasonable to assume that the smaller the number of inputs to a logic element the less complicated – and consequently the cheaper – it will be.

The usual aim of any engineer is to produce a design which completely meets the specification of the system required at the lowest possible cost. Both expressions for R clearly meet the circuit specification but it is probable that the one with the smaller number of elements (and also with elements which have fewer inputs) will be the cheaper one. An additional advantage of the reduced form of circuit is that the smaller the number of components used in any circuit the lower the chance of a component failure when the circuit is in use. Thus the more simple form is probably the more reliable one, and it should be cheaper to maintain.

To achieve minimum cost the circuit designer has to solve a minimization problem. It is assumed here (and in most other treatments of logic circuit design) that the cheapest circuit is the one which requires the smallest number of elements. Therefore minimal cost is regarded as equivalent to producing the minimum form of Boolean expression for the circuit. In a few special cases the minimum form of logic expression is not the cheapest one to implement, but in such cases the minimum expression is rarely much more expensive than the cheapest form.

After producing the minterm form of Boolean expression for a circuit the next stage of circuit design is to reduce this expression to a minimum form. It was indicated in Chapter 2 that the reduction of Boolean expressions by inspection, using the rules of Boolean algebra, is not obvious in many cases. Some method is required which will assist with this reduction; when five or fewer variables are involved one of the most useful techniques is that which uses a Karnaugh map.

3.5   Karnaugh maps

Karnaugh maps are a modification of Venn diagrams, which are a pictorial device that should be familiar to any reader with some elementary knowledge of set theory. It is not essential to understand the origins of Karnaugh maps in order to use them to simplify logical expressions; some of the more advanced texts listed in the bibliography include details of the basic theory of the technique.

A Karnaugh map consists of a rectangular area which is divided into squares (or elements) and each square represents one minterm. There is only one square for any minterm and there is a square for every minterm. The squares are not allocated to the minterms at random, they are arranged so that a movement of one square vertically (up or down) or one square horizontally (left or right) results in the minterms associated with the two adjacent squares differing only in a single variable. In other words the two minterms are identical except for one variable which is inverted in one of them but not in the other. Diagonal movements on the Karnaugh map are not of interest. Within these rules many different maps may be drawn. Figure 3.1 shows two different maps for a system with three inputs of A, B and C; both maps are correct.

A Karnaugh map is best regarded as a three-dimensional device which has to be represented in a two-dimensional form when it is printed. This is one reason why so many maps may be drawn. When movements of one square are repeatedly made in the same direction the edge of the map is eventually reached. A further move of one square is equivalent to moving off the edge of the map and returning onto it at the opposite edge, i.e. the top edge should be joined to the bottom one and the left-hand edge should be joined to the right-hand one. The only way in which the map could be drawn with this

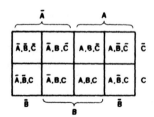
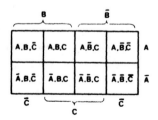

Fig. 3.1   Karnaugh maps

form is on a doughnut-shaped surface (the surface of a toroid). The two-dimensional map is an attempt to represent this surface and it is satisfactory provided that it is remembered that a movement off one edge of the map and back on at the opposite edge is identical to a move from one square to an adjacent square.

Figure 3.2 shows examples of 2-, 3- and 4-variable maps with two different methods of labelling each map. In most cases only a few of the individual squares are labelled with minterms; the form of labelling of each row and column enables the minterm for any square to be determined. Some readers will find one of the forms of labelling easier to use than the other. Usually the form on the right-hand side of Fig. 3.2 is easier to use when inserting logic values from a truth table into the map, while the form on the left of Fig. 3.2 is better when the completed map is used for logic expression reduction. Figure 3.3 is one form of 5-variable map and consists of two 4-variable maps which should be imagined to be placed one above the other.

For systems which have more than five variables, the Karnaugh map technique is not very successful and alternative methods are required. (With care 6-variable maps consisting of four 4-variable maps placed one above the other can be used.) However a very large number of circuits involve five or fewer variables, or may be reduced to several such circuiits, so methods of logic expression minimization based on Karnaugh maps may

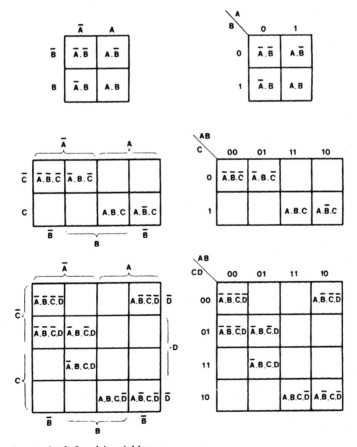

Fig. 3.2 Karnaugh maps for 2, 3 and 4 variables

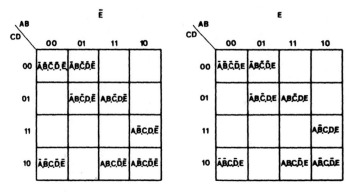

Fig. 3.3 Karnaugh map for 5 variables

be applied to many circuit design problems. In cases for which the maps cannot be used other methods, for example those developed by Quine and McCluskey, are required and are explained in more advanced texts.

3.6 Minimization using Karnaugh maps

To produce a minimal form of Boolean expression with the aid of a Karnaugh map the map must be completed by entering either 1 or 0 in *every* square of the map. Each square corresponds to one minterm (i.e. to one row of the truth table) and 1 is put in those squares which correspond to the cases for which the circuit output is required to be 1; 0 is entered when the output is to be 0. If a circuit has several outputs a separate map should be drawn for each output.

With practice a Karnaugh map may be completed directly from the truth table, but initially the sum of products expression should be formed and used to complete the map.

In Section 3.3 the expression

$$R = A.\bar{B}.\bar{C} + A.B.\bar{C} + \bar{A}.\bar{B}.C + A.B.C$$

was obtained for Example 3.2. The Karnaugh map for R is completed by writing 1 in each square corresponding to the minterms in the expression for R and writing 0 in all the other squares; the completed map for R is shown in Fig. 3.4.

If two vertically or horizontally adjacent squares in the map both contain 1s then it is clear that the map construction is such that only a single variable changes between the two squares. It is simple to show that this variable that changes can be eliminated. For

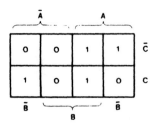

Fig. 3.4 A Karnaugh map for $R = A.\bar{B}.\bar{C} + A.B.\bar{C} + \bar{A}.\bar{B}.C + A.B.C$

example, in Fig. 3.4 the squares for $A.B.\bar{C}$ and $A.\bar{B}.\bar{C}$ are adjacent and both contain 1s. In this case B is the variable that changes and together the squares represent $A.B.\bar{C} + A.\bar{B}.\bar{C}$ which may be rewritten as $A.\bar{C}.(B+\bar{B})$ by use of the Distributive laws of Boolean algebra. The Complement relation $B+\bar{B}=1$ and the Unit rule $(A.\bar{C}).1 = A.\bar{C}$ may be used to further reduce $A.\bar{C}.(B+\bar{B})$ to $A.\bar{C}$. Thus combining two adjacent squares which both contain 1s allows them to be represented by a single AND term instead of by two. Also, the number of variables in the single term is one less than in each of the original two terms.

This grouping together of adjacent squares which both contain 1s is the method by which a sum of products expression is simplified with the aid of a Karnaugh map. The technique can be extended by grouping together adjacent, but not overlapping, groups of two squares to form larger groups. These larger groups *must* be square or rectangular and must have sides which are 2^a squares by 2^b squares where a and b are zero or positive integers. Therefore, on a four-variable map, the groups may be a single square, two squares (2 by 1), four squares (either 2 by 2 or 4 by 1), eight squares (4 by 2) or the complete map. Five- and six-variable maps may have three-dimensional groups of 2^a by 2^b by 2^c where c is also zero or integral.

Each group of 1s formed is equivalent to replacing all the minterms represented by the squares in the group by a single AND expression which contains only those variables which are the same in every square in the group. Thus all those variables which change are eliminated, and the larger the group the greater the degree of simplification. If the whole map contains 1s then all the variables change and the circuit output is 1 regardless of the input conditions.

To produce a minimum logic expression using a Karnaugh map for 2, 3 or 4 variables, all the 1s must be included in groups which are 2^a by 2^b and the groups must be as large as possible. When constructing the groups *it is essential to remember that the maps are global*, i.e. a movement off one edge and back on to the map at the opposite edge is exactly the same as a movement between adjacent squares. Figure 3.5 illustrates a range of different groups on some four-variable maps; some of the groups which run off the edges are not obvious.

When groups are formed every 1 must be included in at least one group but may be included in more than one group if this allows a greater reduction of the whole minterm expression. In other words it is not essential to include a 1 in every possible group which could be formed to include it, but when it is already in one group it may be included in

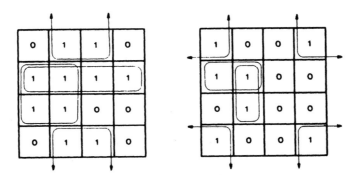

Fig. 3.5 Formation of groups on Karnaugh maps

another if this allows a small group to be replaced by a larger one. For the greatest degree of minimization to be achieved the smallest possible number of groups should be formed.

The following rules summarize the method which should be adopted to produce a minimum logic expression with a Karnaugh map.

a) Form the *largest* possible groups (group sizes may only be 1, 2, 4, 8, $\cdots$ squares)

b) Construct the smallest possible number of groups provided that rule (c) is obeyed.

c) All the squares which contain a 1 must be included in at least one group.

d) Squares should not be included in more than one group *unless* the inclusion of a square in more than one group enables a small group to be replaced by a larger one.

Inexperienced designers commonly make the mistake of using too many groups or constructing groups which are too small. The second fault is most frequently found when groups run off the edges of the map or when groups overlap. Mistakes such as these do not produce a faulty circuit; the circuit operates correctly but it is not the simplest possible. The mistake of attempting to form a group of incorrect size (a size other than $2^a \times 2^b$) will result in a group which cannot be correctly labelled. Any label given to such a group will be wrong and circuits developed using a Boolean expression containing the label will not function as required.

Example 3.3

The Karnaugh maps in Fig. 3.6 illustrate the required behaviour of two logic circuits. Derive minimum logic expressions for both circuits.

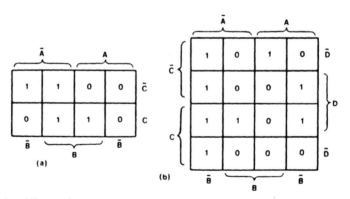

Fig. 3.6 Examples of Karnaugh maps

Solution

a) Figure 3.7 shows the best groupings on the map in Fig. 3.6a. Note that only two groups, each of two squares, are required. A common error on a map such as this one is to form an additional group using the two 1s in the vertical column; this contradicts rules (b) and (d). The upper group on the map is the one in which A is always $\bar{A}$, C is always $\bar{C}$, and B changes, i.e. the group represents $\bar{A}.\bar{C}$. Similar examination shows that the other group is $B.C$ and therefore the minimum logic expression for the circuit represented by the map is $\bar{A}.\bar{C}+B.C$.

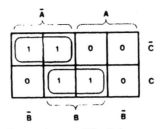

Fig. 3.7 Formation of groups on the Karnaugh map of Fig. 3.6a

b) Figure 3.8 indicates the groups for the map of Fig. 3.6b. In this scheme the square $\bar{A}.\bar{B}.C.D$ has been included in three different groups. If the column of four squares is considered to be an obvious first group, then the remaining four 1s must be included in additional groups. The two 1s on the right-hand edge could be taken as a two-by-one group; however if the two 1s on the opposite edge which have already been used in the column group are also included, a larger two-by-two group is obtained. Similarly, the square $\bar{A}.B.C.D$ can be grouped with the square $\bar{A}.\bar{B}.C.D$ which has already been used twice; now only the square $A.B.\bar{C}.\bar{D}$ remains to be included but as there are no squares containing a 1 adjacent to it then the only group possible is the single square itself. The solution consists of four groups: the column $\bar{A}.\bar{B}$; the group $\bar{B}.D$ which runs off the edges; the group of two $\bar{A}.C.D$; and the single square $A.B.\bar{C}.\bar{D}$. Therefore the minimized expresssion for the circuit is

$$\bar{A}.\bar{B} + \bar{B}.D + \bar{A}.C.D + A.B.\bar{C}.\bar{D}$$

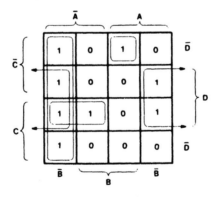

Fig. 3.8 Formation of groups on the Karnaugh map of Fig. 3.6b

3.7 Circuit implementation

The minimum expression which is obtained using the reduction techniques described here is another sum of products form and the circuit could be constructed using several AND gates and one OR gate. However, the principal objective of the minimization technique was the reduction in the cost of the final circuit. If the circuit can be constructed from a single type of gate then bulk purchase of the gates will probably reduce the cost; also, in some electronic logic systems, NAND or NOR gates are the most easily constructed types and are therefore the cheapest.

The lowest cost is usually achieved when a circuit is constructed entirely from NAND gates or entirely from NOR gates (an inverter is regarded as either a single input NAND gate or a single input NOR gate). The sum of products expression obtained from a Karnaugh map minimization is easily converted into either an expression of NAND functions or one of NOR functions by application of de Morgan's theorem.

Example 3.4

Convert the expression for the output X of a circuit into both NAND and NOR function forms. X is the result obtained in part (b) of Example 3.3, i.e.

$$X = \bar{A}.\bar{B} + \bar{B}.D + \bar{A}.C.D + A.B.\bar{C}.\bar{D}$$

Solution

(a) Invert X twice

$$X = \bar{\bar{X}} = \overline{\overline{\bar{A}.\bar{B} + \bar{B}.D + \bar{A}.C.D + A.B.\bar{C}.\bar{D}}}$$

Apply de Morgan's theorem to $\bar{X}$, that is to the first inversion. The result is

$$X = \overline{(\overline{\bar{A}.\bar{B}}).(\overline{\bar{B}.D}).(\overline{\bar{A}.C.D}).(\overline{A.B.\bar{C}.\bar{D}})}$$

which is entirely formed from NAND functions as required.

(b) To obtain the NOR form, each AND term is separately inverted twice giving

$$X = \overline{\overline{\bar{A}.\bar{B}}} + \overline{\overline{\bar{B}.D}} + \overline{\overline{\bar{A}.C.D}} + \overline{\overline{A.B.\bar{C}.\bar{D}}}$$

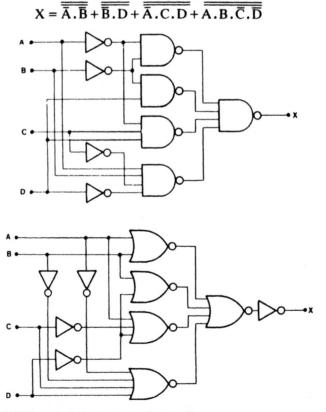

Fig. 3.9   NAND and NOR gate logic circuits for the function X

Again de Morgan's theorem is used and the result is

$$X = \overline{A+B} + \overline{B+\bar{D}} + \overline{A+\bar{C}+\bar{D}} + \overline{\bar{A}+\bar{B}+C+D}$$

Double inversion of this expression gives one which is entirely in NOR functions.

Diagrams for both forms of the circuit are drawn directly from the expressions for X and are shown in Fig. 3.9.

3.8 Unspecified states

In the preceding sections it has been assumed that for every possible input condition the circuit specification explicitly requires a particular output from the circuit. However in some applications of logic circuits it is impossible for certain input conditions to occur and in these cases the circuit output will not be specified; the following is a simple example of one such situation.

An industrial process requires that a section of the production plant is kept between two temperatures. If the temperature exceeds some upper limit a refrigerator is switched on, and if the temperature falls below a lower limit a heater is switched on. Two thermostatic switches, A and B, provide an indication of the temperature to a control unit. Switch A gives a signal of 1 when the temperature exceeds the upper limit and 0 when it does not, while switch B gives 1 when the temprature is too low and 0 otherwise. The control unit is a combinational logic circuit; the signals from A and B are the inputs, and there arc two outputs, R and H. Output R controls the refrigerator so that when R = 1 the refrigerator is on and when R = 0 it is off. Similarly when H = 1 the heater is on and when H = 0 it is off. Table 3.5 is the truth table for this logic circuit.

Table 3.5

Inputs		Outputs		Comments
A	B	R	H	
0	0	0	0	Temperature is within required limits
1	0	1	0	Temperature too high, cooling required
0	1	0	1	Temperature too low, heating required
1	1	?	?	Impossible input condition; would imply temperature above upper and below lower limits.

If a Karnaugh map is used to produce a minimized logic expression how can cases such as A = B = 1 in the temperature controller be taken into account? As this condition can never arise no particular outputs are required. However, to use the Karnaugh map it is necessary to make an entry in every square so the outputs must be chosen to have a value in this case. It does not matter if the value chosen is 1 or 0. Because the circuit operation is not important in this instance these cases are known as **don't care conditions** or just as **'don't cares'**. A second type of 'don't care' situation can arise and will be introduced in Chapter 6; although the origin of the case is different the treatment is identical.

The aim of logic expression minimization was to reduce the circuit complexity and cost. 'Don't cares' can be used to assist in minimization by choosing the circuit outputs in 'don't care' cases to have the values which give the most simple circuit. When 'don't cares' are used in this way they must be indicated in the truth table and on the Karnaugh map by some symbol; the letter X is used here. The minimization technique itself is modified by choosing a 'don't care' condition to give an output of 1 when this reduces the final logic expression and to give an output of 0 in all other cases. The rules for Karnaugh map minimization remain the same with the additional requirement that Xs are to be included in groups of 1s when this increases the group size or reduces the number of groups, the Xs must not be included in groups in other cases. Thus all the 1s must be included in groups but the inclusion of Xs depends on their position relative to 1s.

Note that in the temperature controller example the case $A = B = 1$ might arise if one detector develops a fault. A third circuit output could be provided to switch on an alarm in this case; this addition would not affect the design of the circuits which provide the other outputs.

Example 3.5

The decimal digits 0 to 9 are represented by an 8421 BCD code (see Section 1.5). Derive a Boolean expression for a logic circuit which will produce an output of 1 when any code representing a digit which is an integral multiple of three is input to the circuit.

Solution

As four Boolean variables with sixteen possible combinations are used to represent the ten decimal digits, six-input combinations will never arise and may be considered to be 'don't care' situations. The truth table is easily constructed and is Table 3.6. The table

Table 3.6

Decimal digit	BCD code D	C	B	A	Output R	Minterm
0	0	0	0	0	1	$\bar{A}.\bar{B}.\bar{C}.\bar{D}$
1	0	0	0	1	0	$A.\bar{B}.\bar{C}.\bar{D}$
2	0	0	1	0	0	$\bar{A}.B.\bar{C}.\bar{D}$
3	0	0	1	1	1	$A.B.\bar{C}.\bar{D}$
4	0	1	0	0	0	$\bar{A}.\bar{B}.C.\bar{D}$
5	0	1	0	1	0	$A.\bar{B}.C.\bar{D}$
6	0	1	1	0	1	$\bar{A}.B.C.\bar{D}$
7	0	1	1	1	0	$A.B.C.\bar{D}$
8	1	0	0	0	0	$\bar{A}.\bar{B}.\bar{C}.D$
9	1	0	0	1	1	$A.\bar{B}.\bar{C}.D$
Unused	1	0	1	0	X	$\bar{A}.B.\bar{C}.D$
Unused	1	0	1	1	X	$A.B.\bar{C}.D$
Unused	1	1	0	0	X	$\bar{A}.\bar{B}.C.D$
Unused	1	1	0	1	X	$A.\bar{B}.C.D$
Unused	1	1	1	0	X	$\bar{A}.B.C.D$
Unused	1	1	1	1	X	$A.B.C.D$

shows that four minterms correspond to outputs of 1, and six to 'don't care' cases. Figure 3.10 is the Karnaugh map derived from the truth table with the groups indicated; these groups correspond to the expression for the output which is

$$R = A.D + A.B.\bar{C} + \bar{A}.B.C + \bar{A}.\bar{B}.\bar{C}.\bar{D}.$$

The result in Example 3.5 is a simpler one than that which would have been obtained

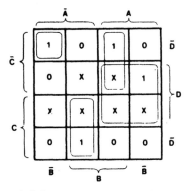

Fig. 3.10   Karnaugh map for Example 3.5

if the six unspecified cases had been chosen to give all 1s or all 0s as outputs. In the groups selected there are four Xs; the effect of including any X in a group is to change the X into a 1, and any X which is not in a group becomes a 0. If these changes are made the map in Fig. 3.11 is obtained. This map fully describes the circuit which has been designed and indicates the output this circuit would produce if an unallowed 'don't care' input condition existed (for example during testing).

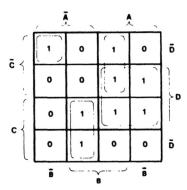

Fig. 3.11   Karnagh map equivalent to the solution of Example 3.5

3.9   Summary of the design method

The design procedure described procedures a reliable and economical combinational logic circuit if correctly applied; it may be summarized as follows.

a)   Obtain a precise circuit specification.

b)   Convert the specification into a truth table which shows the outputs required for every combination of input conditions; 'don't cares' are used for the outputs when the input conditions can never exist.

c) Identify the minterms corresponding to each row in the table.

d) Draw and complete the Karnaugh map for the system.

e) Select groups on the completed maps; form the largest possible groups and the smallest number of groups (of allowed group dimensions).

f) Determine the sum of products expression for the selected groups.

g) Convert the sum of products expression into the one which is most easily implemented using the available components.

h) Draw the circuit diagram.

i) Construct and test a prototype circuit.

3.10 Particular maps

Sometimes the Karnaugh map produced when a circuit is designed is similar to one of the maps in Fig. 3.12. These maps resemble a chessboard with a 1 in the position of every black square and a 0 in the position of every white square. Using the usual simplification rules no groups can be formed on these maps. However, these maps arise when the circuit can be formed very simply using exclusive-OR gates. For example, the map on the left in Fig. 3.12 represents the function $A \oplus B \oplus C = (A \oplus B) \oplus C = A \oplus (B \oplus C)$; this may be proved by deriving the Karnaugh map for the function.

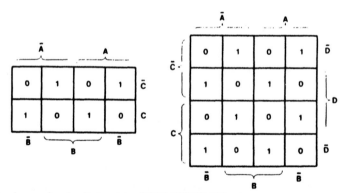

Fig. 3.12   Karnaugh maps for circuits involving EXCLUSIVE–OR gates

3.11 Multiple-output circuits

It has been assumed that multiple-output circuits are designed as a number of independent single-output circuits. This is reliable, but techniques exist which enable more economical multiple-output designs to be produced. These techniques generally require that the user has some experience of logic circuit design; a reasonable compromise for the beginner is to design separate single-output circuits and then examine the minimum expressions carefully. If these minimized expressions are inspected for terms common to more than one expression then the common terms need only be generated once within the complete multiple-output circuit.

Example 3.6

Draw the circuit diagram in NAND gate form for a circuit with three inputs A, B and C and three outputs X, Y and Z. The outputs are given (in minimum form) by

$$X = A.B.\bar{C} + A.\bar{B}.C$$

$$Y = \bar{A}.B + A.\bar{B}.C$$

$$Z = \bar{A}.B + B.\bar{C}$$

Solution

It is apparent that some of the product terms are common to more than one output expression; if $L = \overline{\bar{A}.B}$ and $M = \overline{A.\bar{B}.C}$ then the outputs are given by

$$X = A.B.\bar{C} + \bar{M}$$

$$Y = \bar{L} + \bar{M}$$

$$Z = \bar{L} + B.\bar{C}$$

Applying de Morgan's theorem gives the NAND forms

$$X = \overline{\overline{(A.B.\bar{C})}.M}$$

$$Y = \overline{L.M}$$

$$Z = \overline{L.\overline{(B.\bar{C})}}$$

L and M may be generated using inverters for $\bar{A}$ and $\bar{B}$ and one NAND gate for each. The outputs of these NAND gates are used as inputs to the circuits which generate the final outputs. The complete circuit is shown in Fig. 3.13. Some degree of economy can be achieved in this way but methods which give greater economy do exist.

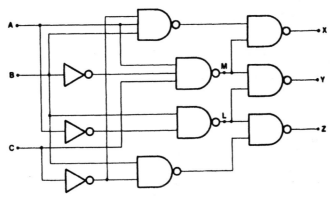

Fig. 3.13   Example of a multiple output logic circuit

3.12   Comment

A single circuit design technique has been described, it is one of many that have been developed. The technique produces an economical circuit design in nearly all cases, this circuit is a reliable one which completely meets the initial specification. There are a few exceptional cases in which the minimal solution is not obtained, but even in these cases the circuit is reliable. Many other techniques exist and should be examined by the reader when familiar with this one.

3.13   Problems

1   Derive a truth table for the circuit whose output R is given by

$$R = A.\bar{B} + B.(C + A.D)$$

2   A piece of equipment has four fault detectors and these are connected into an alarm control unit which is a logic circuit. A single fault is regarded as unimportant, it probably arises from a faulty detector. Devise the truth table for the control circuit which indicates the presence of a fault by the state of the output. (Faults are therefore indicated only if two or more inputs – i.e. detectors – show a fault condition.)

3   Use the Karnaugh map technique to reduce the following expressions to a more simple form.

a) $A.C + A.B.\bar{C} + A$

b) $A.B.C + A.C.D + \bar{B}.C.D + \bar{A}.B.\bar{C}.D + \bar{A}.B.\bar{C}.\bar{D}$

c) $R.T + \bar{R}.S.T + \bar{R}.\bar{S}.T$

d) $L.M.N.P + L.M.\bar{P} + L.\bar{M}.N + \bar{L}.M.\bar{P} + L.N + L.\bar{M}.N.P$

e) $\overline{x.\bar{y}.z}. (\bar{x} + y + z) + \overline{x.y}$

4   Two binary bumbers, *a* and *b*, each of two digits are represented as $A_1 A_0$ and $B_1 B_0$ where A_1, A_0, B_0 B_1 are Boolean variables which form the digits of the binary numbers. Devise a combinational logic circuit, or circuits, to provide outputs of X, Y and Z such that X = 1 when $a > b$, Y = 1 when $a = b$, and Z = 1 when $a < b$.

5   Design a logic circuit which will operate as the control circuit of the alarm system of Problem 2. Derive minimal NAND and NOR gate versions of the circuit.

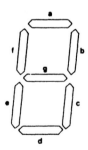

Fig. 3.14   Seven-segment indicator layout

6 Figure 3.14 shows the layout of the now familiar seven-segment indicator. Each segment or bar of such an indicator can be illuminated by applying a logic 1 to the input for that segment; by simultaneously illuminating the appropriate segments, the device can be used to display the digits 0 to 9 in a stylized form.

Design a logic circuit, or circuits, whose seven outputs drive a seven-segment display. The circuit has four inputs; the logic levels at these inputs represent the ten decimal digits in an 8421 BCD code.

7 Example 3.1 describes the action of a circuit to provide a three-way lamp control. Construct a truth table for a four-way lamp control (i.e. four switches controlling a lamp such that a change in position of a single switch changes the on–off state of the lamp). Draw a Karnaugh map for the circuit and comment on the form of the map.

8 The inputs to an automobile safety system are supplied by five on–off switches. Two switches indicate if either door is open, each front seat belt has a switch indicating if it is fastened, and a pressure switch shows if the front passenger seat is occupied. The circuit output is a logic signal which only allows the driver to operate the starter motor when it is safe to do so (e.g. both doors closed). Design a minimal combinational logic circuit to implement the safety system.

9 The four inputs to a circuit represent the ten decimal digits in an 8421 weighted BCD code. Devise a circuit with four outputs which converts the code at the inputs to a different code at the outputs. The output code is a 5421 weighted BCD code.

10 A five-digit binary (5-bit) result is obtained if two 8421 weighted BCD numbers are added together as if they are simple binary numbers. This result can only be in the range zero to eighteen.

Design a circuit which has this 5-bit number as its inputs. Four of the circuit outputs form an 8421 weighted BCD number equal to the 5-bit input when this is less than ten. When the 5-bit input exceeds nine the four outputs represent a value ten less than the input value. A fifth output is 0 when the 5-bit input was less than ten and is 1 when it was greater than nine.

4 Sequential logic elements

The logic circuits described in the preceding chapters are such that the outputs specified by the truth table appear as soon as the inputs are connected. In any real circuit there will be a small time interval between the connection of inputs and the appearance of the outputs because the circuit takes a finite time to operate. This interval is called **propagation delay** and has no effect on the final output(s) of the circuits. One important feature of the combinational logic circuits which were examined earlier is that output states previously held by such circuits have no effect on their present behaviour.

When the output of a circuit depends on past inputs (and hence on existing or previous outputs) as well as the present inputs the circuit action must be considered as a function of time. Any logic circuit in which the order in time of applying inputs is important is termed a **sequential circuit**, i.e. the inputs must follow a specific sequence to produce a required output. In order to follow a sequence of inputs the the circuit must contain some form of memory to retain knowledge of those inputs which have already occurred. This memory is usually obtained by feedback connections which are made so that the effect of the earlier inputs is maintained.

Sequential logic systems are usually divided into two groups, synchronous and asynchronous circuits. A synchronous circuit or system is one in which all the changes take place simultaneously at a time determined by a signal at some control input common to all sections. In an asynchronous system there is no common control; a change in one section of the system causes further changes in other sections and so on. The changes propagate through the system in a manner which is determined only by the speed with which each section operates.

Most sequential systems are based on a small number of simple sequential circuit elements known as **bistables** or **flip-flops**, so-called because they have two stable conditions and can be switched from one to the other by appropriate inputs. These stable conditions are usually called the **states** of the circuit.

4.1 The set–reset flip-flop

This is the most simple sequential circuit element and is commonly referred to by its initials as an SR flip-flop. The logic circuit may be constructed in many ways; Fig. 4.1a is one form using NAND gates and Fig. 4.1b is another using NOR gates.

The circuit has two inputs, S and R, and two outputs, Q and Q'. The feedback mechanism required to form a sequential circuit can be clearly seen; each of the outputs is connected as one of the inputs to a gate which controls the other output. Unlike the combinational logic circuits considered previously the circuit outputs depend on the inputs *and also on the outputs* – they are not just functions of the inputs.

Examination of Fig. 4.1a shows that the following relationships hold.

$$Q = \overline{S.Q'} \quad \text{and} \quad Q' = \overline{R.Q}$$

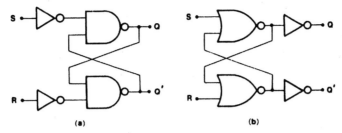

Fig. 4.1   The set-reset (SR) flip-flop

Application of de Morgan's theorem converts these into

$$Q = S + \bar{Q}' \qquad \text{and} \qquad Q' = R + \bar{Q}$$

Similarly the circuit of Fig. 4.1b obeys the equations

$$Q = \overline{\overline{S + \bar{Q}'}} \qquad \text{and} \qquad Q' = \overline{\overline{R + \bar{Q}}}$$

which can easily be reduced to the equations obtained for the circuit of Fig. 4.1a.

An analytical treatment of this type of circuit is not simple and instead of a detailed analysis the circuit is examined in all four possible cases.

a) **S = R = 0**   This is the normal rest state of the circuit and all the equations show that Q and Q' are different, i.e. $Q = \bar{Q}'$. There are two possible ways in which Q and Q' may differ; either Q = 1 with Q' = 0, or Q = 0 with Q' = 1. Both cases are stable ones; the term **stable** means that the circuit will remain in one state and will not change as long as the input conditions remain fixed, i.e. while S = R = 0. However the circuit equations do not specify which state the circuit is in, this depends upon the previous history of the circuit. Thus, unlike a combinational logic circuit, the circuit equations of a sequential logic circuit do not always give a complete description of the outputs.

b) **S = 1, R = 0**   The circuit equations give $Q = 1 + \bar{Q}'$ for which the result is Q = 1 regardless of the value of Q'. Consequently $Q' = 0 + \bar{Q}$ gives $Q' = 0 + 0 = 0$. Hence these input conditions force Q to become 1. Any action which ensures that Q = 1 is a **SET** action, and this is why the S or SET input is so-called.

c) **S = 0, R = 1**   This is the exact opposite of case (b) and the circuit outputs are Q = 0 and Q' = 1. An action causing Q to become 0 is a **RESET** action which is why the second input is the R or RESET input.

d) **S = R = 1**   This is the most difficult case; the circuit equations give Q = Q' = 1 which results in a situation unlike the previous three cases where Q and Q' are always different. The situation is stable while S = R = 1 but a problem arises if S and R change simultaneously from 1 to 0. Examination of the circuit equations suggests that as the inputs becomes S = R = 0 the circuit should go to either of the possible $Q = \bar{Q}'$ states, but there is no information to indicate which one. A detailed examination of the operation of the individual gates in the circuit of Fig. 4.1a shows that if S and R go to 0 simultaneously while Q = Q' = 1 then both Q and Q' should become 0. However, if Q = Q' = 0 while S = R = 0 both Q and Q' should go to the 1 state and so on. Therefore, the circuit should become unstable and oscillate between the states Q = Q' = 1 and Q = Q' = 0.

In any real circuit one of the output gates will operate marginally faster than the other and the circuit will not oscillate but will go to one of the $Q = \bar{Q}'$ states; which one of the two states cannot be predicted. In this particular case the circuit action is indeterminate and a **race condition** or **hazard** is said to exist. Many different types of hazard may arise in logic circuits; the term is used whenever the output of a circuit cannot be predicted or when an incorrect output is produced (even if only for a brief time) before the circuit produces the correct one. A race condition is one particular type of hazard and arises when the final circuit state depends upon the relative operating speeds of two or more circuit components. Hazard conditions may occur in any sequential logic circuit; they must be identified and excluded in some way.

When circuits are constructed incorporating SR flip-flops the design should be such that the condition $S = R = 1$ never arises. In these circumstances, Q and Q' are always different and the outputs may be labelled Q and $\bar{Q}$.

4.2   The operation and use of an SR flip-flop

The use and action of an SR flip-flop can be described by the following statements.

    a) S and R are normally held at 0 and the outputs remain constant in either one of the $Q = \bar{Q}'$ states.

    b) An input sequence of 0 to 1 then back to 0 at the S input will ensure that $Q = 1$ and $Q' = 0$.

    c) A similar 0–1–0 input sequence at the R input ensures that $Q = 0$ and $Q' = 1$.

    d) In normal circuit design the input condition $S = R = 1$ should not be allowed.

    e) If power is connected to the circuit with $S = R = 0$, the circuit will take either one of the states $Q = \bar{Q}'$.

If an SR bistable is operated so that only these conditions may exist then the Q' output is always the inverse of the Q output. Therefore the Q' output is usually labelled $\bar{Q}$; this is the nomenclature most often used for commercial SR devices.

Applications which require that a bistable takes some specific initial state require additional circuits to force Q to the starting value. Such initialization requirements are common in large systems and the additional setting circuits are often operated by a manual start switch. This start switch also functions as a manual system reset.

Logic circuit diagrams of large networks, particularly those constructed from transistor-transistor logic (TTL) elements, often include the circuit of Fig. 4.2. This is sometimes wrongly called an SR flip-flop; it is a $\bar{S}\,\bar{R}$ flip-flop (a not S, not R flip-flop). The circuit has definite set and reset actions but all the input and operating conditions for a $\bar{S}\,\bar{R}$ are the inverse of those for an SR device. For a $\bar{S}\,\bar{R}$ bistable the input conditions $S = R = 0$ should be avoided, and set and reset actions require 1–0–1 transitions at the inputs. The $\bar{S}\,\bar{R}$ circuit has many uses; a common application is as a buffer to overcome problems which occur when mechanical switches are connected to electronic logic systems (see Section 7.4).

Another logic circuit is shown in Fig. 4.3. This one is constructed from NOR gates. The only difference between this circuit and those of Fig. 4.1 is that in the case $S = R = 1$ the circuit has the output $Q = Q' = 0$ instead of $Q = Q' = 1$. If the $S = R = 1$ input

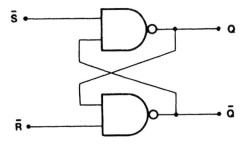

Fig. 4.2 The $\bar{S}\bar{R}$ flip-flop

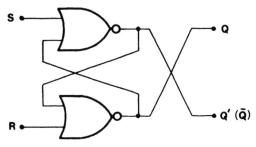

Fig. 4.3 An alternative logic circuit for an SR flip-flop

condition is not allowed then the circuit behaviour cannot be distinguished from that of the circuits of Fig. 4.1.

Bistables may be represented by networks of basic logic gates showing how they are constructed. However bistables are widely used components and there are standard symbols available to represent them. In general the symbols for bistables appear to be simple labelled boxes; however they have many subtle features which necessitate care in their use. For example there are five slightly different symbols for SR bistables, these cover all the different possible circuit actions for the $S = R = 1$ condition. When more complicated bistables are used the range of symbols is even greater. Symbols for bistables are not described in detail in this book; a few symbols for specific cases are used when required to enable circuits to be drawn in a compact form.

4.3 The clocked SR flip-flop

Figure 4.4 is the circuit diagram of an SR flip-flop which has an additional input

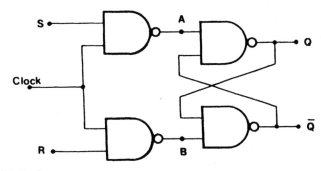

Fig. 4.4 Clocked SR flip-flop

designated clock or enable. This input is used to change the SR flip-flop from an element used in asynchronous sequential circuits to one which may be used in synchronous circuits.

The clock input is the control input which may be used to cause the circuit to operate at a precisely determined time, this is the essential feature of a synchronous system. When the clock input is at logic 0 the points in the circuit diagram labelled A and B are both at 1 regardless of the inputs at S and R. The remainder of the circuit is a $\bar{S}\bar{R}$ flip-flop with A and B as the inputs. Therefore, as long as the clock input remains 0, the inputs to the $\bar{S}\bar{R}$ section stay at 1 and the final outputs remain constant at the present values of Q and $\bar{Q}$. Hence while the clock is at 0 the S and R inputs have no effect on the circuit and may change many times (they may even take the S = R = 1 state) without affecting the outputs. Only when the clock input becomes 1 do S and R affect the output; when the clock is at 1, point A equals $\bar{S}$ and B equals $\bar{R}$, so that while the clock remains 1 the complete circuit behaves exactly as an SR flip-flop without a clock input.

To operate the flip-flop as a synchronous element S and R are set to the required values while the clock input is 0. A 0 to 1 to 0 sequence (called a **clock pulse** or just a **pulse**) is then applied to the clock input with S and R kept constant. The flip-flop operates exactly as the non-clocked version except that any changes in the output states occur at a time determined precisely by the signal at the clock input. If the same clock pulse is applied simultaneously to a number of flip-flops in a large system they will all operate at the same time. Circuits which have this feature of common clock connected to all bistables in the circuit are more easily designed, and have fewer operating problems, than circuits without such a clock.

When a clocked flip-flop is used in this precisely timed manner its behaviour may be expressed in a tabular form. The form is similar to that of the truth tables used to describe combinational logic networks but differs in that the table for a clocked flip-flop shows the time dependence of the circuit. (This table is sometimes called a truth table, particularly in the data sheets supplied by integrated circuit manufacturers; a more suitable name for it is transition table or switching table.) The behaviour of a clocked SR flip-flop is given by Table 4.1.

Table 4.1

S	R	t_n Q_n	$\bar{Q}_n$	t_{n+1} Q_{n+1}	$\bar{Q}_{n+1}$	Comments
0	0	0	1	0	1	No change in outputs
0	0	1	0	1	0	
1	0	0	1	1	0	SET action
1	0	1	0	1	0	
0	1	0	1	0	1	RESET action
0	1	1	0	0	1	
1	1	0	1	?	?	Operation indeterminate
1	1	1	0	?	?	

In the table t_n denotes the time before a clock pulse and t_{n+1} denotes the time after the clock pulse; Q_n and Q_{n+1} are the values of Q which correspond to these times. An alternative notation which is sometimes used is Q_- for Q_n and Q_+ for Q_{n+1}. The comments have been added to the table to emphasize the circuit behaviour and, as indicated, the action in the case $S = R = 1$ cannot be predicted as it is the race condition again.

4.4   The D-type flip-flop (or latch)

If an inverter is connected so that the R input to a clocked SR flip-flop is always the inverse of S then the complete circuit has a single input, D, and is known as a D-type flip-flop. This is shown schematically in Fig. 4.5.

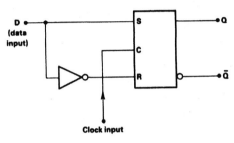

Fig. 4.5   The D-type flip-flop

Because R is always $\overline{S}$ the hazardous condition $S = R = 1$ cannot arise; the complete action of the flip-flop is simpler than that of the SR flip-flop and is given in Table 4.2.

The D-type flip-flop will act as a storage element for a single binary digit (bit). The logic states 1 and 0 may be used to represent the two binary digits 1 and 0 resepctively. If the bit to be stored is presented as the appropriate logic level at the D (data) input and a clock pulse is applied to the flip-flop with the input maintained at D as long as the clock input is at 1, then Q will become the same as D. Thereafter Q will remain at this value until a new binary digit is input in the same way. Therefore a bit input at D is held (stored) by the flip-flop even when it no longer exists at D.

Table 4.2

	t_n		t_{n+1}	
D	Q_n	$\overline{Q}_n$	Q_{n+1}	$\overline{Q}_{n+1}$
0	0	1	0	1
0	1	0	0	1
1	0	1	1	0
1	1	0	1	0

Some integrated circuit D-type bistables do not operate in the manner which has been described. These alternative devices are edge-triggered types and the behaviour of such devices is described later. To distinguish between the two forms of D-type bistable the version already described, that is a simple clocked SR bistable with an inverter between the S and R inputs, is often called a **latch**.

If a number of D-type flip-flops are connected to the same source of clock pulses then the group of flip-flops will store several binary digits which may be used to represent a multiple digit binary number. This group of digits is called a **word** and the group of flip-flops is called a **register**. This grouping can be arranged in many ways and there are many types of register.

4.5   The serial shift register

A serial shift register consists of a number of storage elements (bistables, flip-flops) arranged in a circuit with a single data input, a single clock input, and several outputs which are denoted here as OP_A, OP_B, OP_C, . . . , etc. A new value is set up at the data input while the clock input is at 0. A single clock pulse then causes the new input to appear at the first stage output, OP_A; the original value of OP_A is transferred to OP_B, the original value of OP_B is transferred to OP_C, and so on. Therefore each clock pulse causes the contents of the register to move along (shift) one flip-flop (stage). New data must be input to the first stage and the contents of the last stage are lost. In some special applications the output of the last stage is connected as the first stage input and a recirculating action is produced.

A common (**incorrect**) method of constructing a serial shift register attempted by many new to logic design is illustrated for three stages in Fig. 4.6. The circuit consists of a number of simple clocked SR flip-flops connected in cascade (D-type latches could have been used); three stages are shown. However, when the flip-flops used in Fig. 4.6 are the simple clocked SR type already described the circuit will not operate as required. (It will operate correctly if the flip-flops are one of the more complex types still to be introduced.) When the clock is at 1 the new input appears at output OP_A; for electronic logic gates this will occur very soon after the clock becomes 1 as they have a small propagation delay. Consequently when the new value appears at OP_A the clock will probably still be at 1 and this value will be transferred to OP_B. If the clock is still at 1 when the value appears at OP_B it will be transferred to OP_C, and so on. Thus all the outputs will show the same value, equal to that at the input, unless the clock pulse is of short duration (in this case an unpredictable number of outputs will all show the new input value).

One reliable method of creating a shift register which functions correctly is to double the number of flip-flops and to use two sequences of clock pulses which have a constant relationship to one another in time. The two clock pulse sequences are known as a

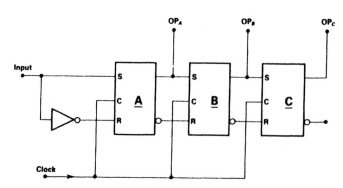

Fig. 4.6   An incorrect circuit for a shift register

two-phase clock and two different sequences are shown in Fig. 4.7. Figure 4.7a shows a general purpose two-phase clock while Fig. 4.7b shows a more simply constructed form in which phase ϕ_2 is just the inverse of phase ϕ_1. The form shown in Fig. 4.7b may give problems in some applications and should only be used if the possible timing problems are understood and avoided.

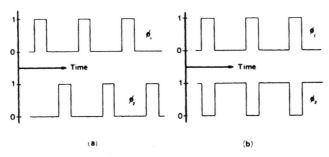

Fig. 4.7 Two phase clock sequences

A modified shift register using two flip-flops for each stage and a two-phase clock is shown in Fig. 4.8; only three stages are illustrated.

This arrangement is called a **master-slave** configuration (or sometimes a double-ranked one); the flip-flops A, B and C are the master flip-flops and A', B' and C' are the slaves. The rest condition is $\phi_1 = 0$ and hence $\phi_2 = 1$, so that the slaves are operating (enabled) but the masters are held inactive and their outputs remain constant. Since the input A' is Q_A the output of A' becomes the same as that of A, i.e. $Q_{A'} = Q_A$. Similarly $Q_{B'} = Q_B$ and $Q_{C'} = Q_C$. When ϕ_1 becomes 1 then ϕ_2 becomes 0, the masters are now active and the slaves inactive. $Q_{A'}$, $Q_{B'}$, $Q_{C'}$ remain at the previous values but Q_A takes the value of the new input, Q_B becomes $Q_{A'}$, and Q_C becomes $Q_{B'}$. When ϕ_1 returns to 0 the masters are again inactive and the slaves active. Now the new states at the Q outputs of the masters are transferred to the slave Q outputs which are also the external outputs. The two sets of flip-flops are clearly connected to operate in an alternating sequence in such a way that the inputs to any flip-flop remain constant while that flip-flop is active. This overcomes the problems associated with the circuit of Fig. 4.6. The action is shown in combined graphical and tabular form in Fig. 4.9. I_4 represents the new input and I_3, I_2 and I_1 represent the initial values of $Q_{A'}$, $Q_{B'}$ and $Q_{C'}$ respectively.

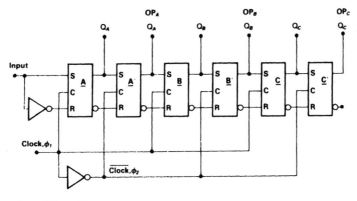

Fig. 4.8 A master-slave shift register

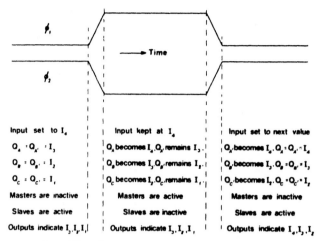

Fig. 4.9   Timing of a master-slave shift register

4.6   Other registers

The shift register described above is one in which data is input one bit at a time to the first stage and the outputs of each stage are obtained at OP_A, OP_B, and OP_C. The outputs of all stages are available simultaneously – i.e. they may be connected to some other circuit in a parallel arrangement, but the inputs must arrive one after the other in a serial order in time along a single connection from the source. This type of register is a serial in-parallel out (SIPO) shift register.

Some integrated circuit shift registers are similar but the user can only obtain the output of the last stage (OP_C in the three-stage example) so that the output may only be obtained one bit at a time. This output is in serial form and the register is a serial in-serial out (SISO) shift register. Obviously the SIPO type may be used as a SISO type but the reverse is not possible.

In addition to being constructed with either serial or parallel output, registers may be constructed with serial or parallel input. A parallel input register is one which may have all the elements (stages) loaded simultaneously from separate (parallel) input connections instead of by repeated input of a single bit combined with a serial shift operation.

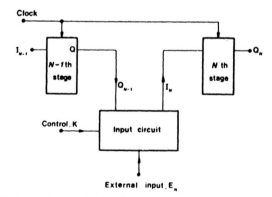

Fig. 4.10   Input to the *N*th stage of a controlled register

Some parallel input registers are also capable of a shift action; the operation of such dual purpose registers depends on the status of a control input at the time at which a clock pulse is input to the register.

The design of registers which have this controlled behaviour is simple; Fig. 4.10 is a block diagram for the Nth stage of such a register and shows a simple register with an additional section labelled 'input circuit'. The action of this input circuit is that if the control, K, is in one logic state then the register stage is loaded from the external input (parallel load) and if the control is in the other state the register shifts, i.e. the $N-1$th stage output is used as the Nth stage input. The control, $N-1$th stage output (Q_{N-1}), and the external input (E_N) form the inputs to the block called 'input circuit'. This block is a combinational logic circuit whose output is the Nth stage input, I_N. If $K=0$ corresponds to the case of parallel loading and $K=1$ corresponds to serial shifting then the input circuit truth table, Table 4.3, can be constructed immediately.

Solution for I_N in terms of K, Q_{N-1} and E_N gives

$$I_N = K.Q_{N-1} + \bar{K}.E_N = \overline{\overline{K.Q_{N-1}}.\overline{\bar{K}.E_N}}$$

One input circuit is required for every stage (except the first) and the control K is connected as a common input to every input circuit.

Table 4.3

Control K	External input, E_N	$N-1$ stage output, Q_{N-1}	N stage input, I_N
0	0	0	0
0	0	1	0
0	1	0	1
0	1	1	1
1	0	0	0
1	0	1	1
1	1	0	0
1	1	1	1

Another feature which can be chosen when a shift register is designed is the shift operation itself. Some parallel in-parallel out (PIPO) registers have no shift operation at all, many registers shift in one direction only while others may be shifted in either direction. The design of registers which shift in either direction is similar to that of registers which can be parallel loaded or serial shifted; essentially the parallel input, E_N, is replaced by the output of the $N+1$th stage, Q_{N+1}.

When the number of stages of a register is specified, three other features must also be given to describe it. The input must be chosen to be serial or parallel (or both with a control); the output may be serial or parallel; there may be no shift operation, single direction shifting only, or shifting may be possible in either direction.

4.7 The JK flip-flop

This is the most complicated of the normally available flip-flops but it is probably the most important. One reason for its importance is that it may be connected to behave as any one of the other types introduced. Furthermore, the $J = K = 1$ condition (equivalent to $S = R = 1$ for an SR flip-flop) is not indeterminate but is defined to give a very useful change-over (toggle) action by which Q_{n+1} becomes $\bar{Q}_n$. The circuit has a clock input and two control inputs, J and K; operation of the circuit is completely described by Table 4.4.

Table 4.4

		t_n		t_{n+1}	
J	K	Q_n	$\bar{Q}_n$	Q_{n+1}	$\bar{Q}_{n+1}$
0	0	0	1	0	1
0	0	1	0	1	0
1	0	0	1	1	0
1	0	1	0	1	0
0	1	0	1	0	1
0	1	1	0	0	1
1	1	0	1	1	0
1	1	1	0	0	1

Although Table 4.4 is a complete description of the behaviour of a JK flip-flop it is often useful to indicate the manner in which the bistable is operating. Table 4.5 describes the operation being performed by the bistable and it is convenient to refer to it as an **action table**.

Table 4.5

J	K	Action at next clock pulse
0	0	No change, $Q_{n+1} = Q_n$
1	0	SET, $Q_{n+1} = 1$
0	1	RESET, $Q_{n+1} = 0$
1	1	Change over (toggle), $Q_{n+1} = \bar{Q}_n$

Two forms of integrated circuit JK flip-flop are available; these are master-slave and edge-triggered versions of the bistable. Master-slave circuits are described first as their internal operation is more easily understood than that of edge-triggered devices. The operation of edge-triggered circuits is examined in Section 4.8.

All bistables may operate spuriously (change output when the necessary input to cause the change does not appear to exist) if incorporated in a network which is poorly designed or constructed. Bad practice in making and decoupling power supply connections, or the use of input signals which do not change quickly and smoothly from one logic level to the other are the main causes of operating problems. Poor input signals are often caused by connections which are long and follow poor routes. Spurious bistable output operations from these, and similar, causes occur more often with edge-triggered devices than with master-slave devices (edge-triggered devices can respond to shorter duration effects). The inexperienced circuit designer should use master-slave devices, if they are available, to reduce the risk of spurious operation.

As in the case of the simple SR flip-flop the JK type may be constructed in many ways. Figure 4.11 is the logic circuit diagram of a master-slave JK flip-flop which uses only NAND gates. The circuit is basically two clocked SR flip-flops connected in a master-slave configuration with the addition of further feedback. This additional feedback consists of cross-coupled connections from the slave outputs to extra inputs on the first gates of the master. There are thus two levels of feedback in this circuit rather than just one level. This makes analysis of the circuit behaviour difficult, but examination shows that there is no indeterminate condition and the useful toggle action is introduced.

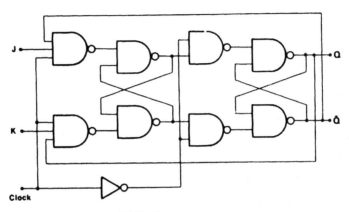

Fig. 4.11   Logic circuit for a master-slave JK flip-flop

If the circuit is examined in all possible circumstances, i.e. for all possible combinations of values of J, K and Q_n, then the behaviour given by Table 4.4 is obtained. This examination is left as an exercise for the reader who is interested in the detailed operation of this particular design of JK flip-flop.

Another way to express the behaviour of a clocked flip-flop is in a Boolean expression for the next output (i.e. Q_{n+1}) in terms of the inputs and the present output. In the case of the JK flip-flop J, K and Q_n are regarded as input variables and Q_{n+1} is the output variable; a Boolean relation is then derived for Q_{n+1} in terms of J, K and Q_n. To produce a minimal form of Boolean expression a Karnaugh map may be constructed as shown in Fig. 4.12; the relationship obtained with the aid of the map is

$$Q_{n+1} = J \cdot \bar{Q}_n + \bar{K} \cdot Q_n$$

A Boolean description can only be produced for those clocked flip-flops which have no indeterminate operations.

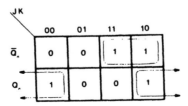

Fig. 4.12   Karnaugh map defining Q_{n+1} for a JK flip-flop

The exact time at which a clocked flip-flop operates, i.e. the time at which the new output appears, depends upon its construction and is different for master-slave and edge-triggered forms. These timing differences require that any logic network should be designed using a single type of clocked bistable. That is master-slave and edge-triggered devices should not be used in the same circuit; if used together there is a high probability of timing hazards which will cause incorrect circuit behaviour.

When master-slave bistables are used the Q and $\bar{Q}$ outputs only change after the clock input returns to 0. The new outputs appear after the full 0–1–0 pulse sequence, until the pulse is complete the outputs remain at their previous values. If J and K change while the clock is at 1 the final circuit action depends upon the internal construction of the flip-flop; this is usually unknown (e.g. the exact internal form of integrated circuit bistables is rarely supplied by manufacturers). Unless detailed specifications for the circuit behaviour are available J and K should be set to the required values while the clock is at 0. These inputs should be maintained constant throughout the full 0–1–0 clock pulse.

Many integrated circuit JK flip-flops have multiple J and K inputs. In most cases, J is defined to be the AND function of all the J inputs. Thus if the inputs are J_1, J_2 and J_3, then the value of J which determines the flip-flop operation is given by $J = J_1.J_2.J_3$. Similarly K is usually the AND function of all the K inputs. However, it is important to note that *this is not always the case* and it is necessary to read the specification of multiple-input devices very carefully. There are also devices with inverted inputs. One common integrated circuit has a normal J input but the other control input is $\bar{K}$.

Another common feature of integrated circuit flip-flops is the provision of CLEAR and PRESET inputs. There is no general convention which specifies the logic level required to activate such inputs; some devices require a 1 while others require a 0. However the action is such that when the appropriate level is applied to the CLEAR input, the Q output immediately becomes 0 regardless of all other inputs. An input to PRESET results in $\bar{Q}$ being forced immediately to the 0 state. In general the condition that Q and $\bar{Q}$ outputs are different is retained, but if both CLEAR and PRESET are activated simultaneously a problem arises; there is no standard circuit behaviour, but in many cases both Q and $\bar{Q}$ are forced to 1 while both inputs are active.

The purpose of the CLEAR and PRESET inputs is not to provide a flip-flop with more modes of operation in normal circuit constructions. They should only be used to initialize circuit conditions; i.e. they are connected so that the user of a complex circuit may force it into a known starting condition. Any other use of the CLEAR and PRESET inputs requires knowledge of the many problems which arise because these control inputs have an immediate effect whereas the other control inputs only have an effect at the time at which the clock input changes.

4.8 Edge-triggered bistables

Several types of bistable have been introduced and most have a clock input which controls the time at which the outputs change. All the bistables already examined require a complete 0–1–0 clock pulse for correct operation, that is they operate using two changes in level at the clock input. Alternative versions of clock synchronized bistables exist which operate with a single change in level at the clock input and are called **edge-triggered** bistables. Those requiring the 0 to 1 clock transition to operate are positive edge-triggered, and those which operate with the 1 to 0 transition are negative edge-triggered.

The transition table for an edge-triggered bistable is exactly the same as that for any other form which has the same control inputs; for example Table 4.4 is the transition table for both master-slave and edge-triggered JK bistables. Edge-triggered bistables and other forms only differ in their behaviour as a function of time, not in their basic function. The control inputs (i.e. S and R, or D, or J and K) of the bistables already described must remain constant throughout the complete clock pulse. Edge-triggered devices only require the controls to be constant during a single transition (edge) of the clock pulse.

A positive edge-triggered JK flip-flop is one example of a readily available edge-triggered bistable. For an ideal device the J and K inputs must be fixed at the required values while the clock input is 0 and held constant until the 0 to 1 clock transition is complete. The Q and $\bar{Q}$ outputs appear as soon as the clock output reaches 1; any subsequent 1 to 0 clock transition has no effect on the bistable. Obviously a negative edge-triggered device requires the control inputs to be set up while the clock input is at 1 and held constant through the 1 to 0 change, the new output appears when the clock has become 0.

For an ideal bistable the control inputs could be fixed at the appropriate values an infinitesimally short time before the clock transition. Any real device requires the inputs to be constant for some time before the clock input changes; this period with constant control inputs is the **set-up time** for the bistable. Most edge-triggered bistables also require clock input signals which change smoothly and very rapidly from one logic level to the other. A final deviation from ideal behaviour is that the bistable does not change state immediately the clock transition is complete. This requires that the control inputs are held constant for a short time, the **hold time**, after the clock transition; also there is a propagation delay between the clock transition and the appearance of the new outputs. The extent to which any device deviates from ideal behaviour depends upon its design and the components used to construct the bistable. Values for one typical integrated circuit bistable are a minimum set-up time of 20 nsec (1 nanosecond = 10^{-9} seconds), a minimum hold time of 5 nsec, and a maximum propagation delay of 40 nsec.

Reliable operation of edge-triggered bistables requires them to have been designed very carefully. If the schematic logic circuits provided by many integrated circuit manufacturers are constructed from basic logic gates they usually do not function correctly. A detailed examination will show that most of these designs include several possible race hazards. The expert designer with control of the speed of individual circuit components can ensure that particular components are always the fastest ones so that the circuit operates correctly. This control of the relative speeds of individual components within a circuit may be achieved accurately by manufacturing the circuit as a single integrated circuit. Consequently, although edge-triggered circuits have existed for a long time, the rapid increase in the proportion of bistables which are edge-triggered reflects developments in integrated circuit manufacture and application.

4.9   Operation of an edge-triggered bistable

There are two distinct steps in the operation of any clock controlled bistable. First the control inputs are allowed to cause changes in the state of the bistable, some time later these changes are 'locked-in' allowing the control inputs to be removed without their removal affecting the bistable. This two stage process has already been described for one design of master-slave bistable with the two steps occurring at times set by the two transitions of the clock pulse.

It appears impossible for a two-stage process to be controlled by a single clock transition as required in the operation of an edge-triggered bistable. Such devices function by use of propagation delays, the single clock signal travels along two routes with different delays arranged so that a specified route is always the faster one. Therefore the signal arrives at two points within the device at different times and can cause a two-stage process to take place. Essentially the circuit design deliberately includes a race hazard (a feature designers normally avoid) and only functions because the circuit components are selected so that the race is always 'won' in the same way. Reliable operation requires expert design of the bistable.

Construction of the deliberate race hazard is illustrated in Fig. 4.13. In the following discussion it is assumed that all the gates have identical propagation delays of τ. That is, the time between an input of a gate changing and any resulting change in output is τ. Also it is assumed that when an input or output does change the change occurs instantaneously, as shown in Fig. 4.13b. In Fig. 4.13a point $B = \bar{A}$ as it is produced by three inversions of A, hence $R = A.B = A.\bar{A} = 0$ using the rules of Boolean algebra. However each of the inverters takes a time to operate. Therefore when A changes from 0 to 1 there is not an immediate change of B from 1 to 0; consequently there is a short period during which $A = B = 1$ giving $R = A.B = 1$. The values of A, B and R are shown as a function of time in Fig. 4.13b with allowance made for the delays in each gate. This illustrates how a single edge of a pulse may be used to create a complete pulse. The three real inverters with delays are a simple form of signal delay circuit which is sometimes called a **delay gate**. In principle the circuit of Fig. 4.13a could be used at the clock input of a master-slave bistable to convert it into an edge-triggered type. The time period marked t_1 would form the set-up time and that marked t_2 would be the hold time. (See Chapter 7 for a further discussion of the effects of propagation delay.)

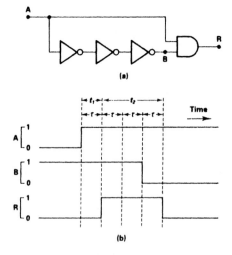

Fig. 4.13 Formation of a pulse from one signal transition

In practice, edge-triggered bistables are not built by combining a circuit of the form of Fig. 4.13a with a master-slave bistable. The delay elements are incorporated into the full bistable design and the deliberate race introduced is not obvious. There are many different designs for edge-triggered bistables. Figure 4.14 is the schematic logic circuit for a negative edge-triggered JK flip-flop; it is similar to the 'typical circuit' suggested by many integrated circuit manufacturers. Any attempt to determine the behaviour of this circuit using only Boolean relationships will not give a correct description of the action. Also, if the circuit is constructed using basic logic gates it will probably not operate correctly (whereas if the master-slave circuit of Fig. 4.11 is built carefully from separate gates it will behave as expected). The circuit only operates correctly when constructed so that the two NAND gates have much greater propagation delays than all the other gates. Determination of the circuit operation must take the gate delays into account.

To demonstrate the circuit behaviour one of the many possible transitions is examined in detail. The situation $J = K = 1$, $Q = 0$, $\bar{Q} = 1$, and Clock = 1 is assumed to have existed for a long time (i.e. the circuit is stable and no changes will occur until some input changes). With these conditions the NAND gate outputs, Y and Y', take the levels $Y = 0$ and $Y' = 1$.

The clock now changes from 1 to 0 and it is assumed that the two NAND gates operate so slowly that Y and Y' remain fixed at $Y = 0$ and $Y' = 1$ until all other changes have occurred. Initially the upper AND-NOR group has the output $Q = \bar{X} + \bar{Y}.Z = \bar{1} + \bar{0}.1 = 0$ and when the clock (Z input) changes to 0 this becomes $Q = \bar{1} + \bar{0}.0 = 1$. This value will remain 1 even if the X input changes; such a change is possible because X is the output of the other AND-NOR group. After Q becomes 1 examination of the lower AND-NOR group gives $\bar{Q} = \bar{X'} + \bar{Y'}.Z' = \bar{1} + \bar{1}.0 = 0$. Thus $\bar{Q}$ becomes 0 and is fed back to the upper group maintaining $Q = 1$ regardless of Y and Z. (The two AND-NOR groups together form an unusual design of $\bar{S}\bar{R}$ flip-flop with inverted clock action.)

Consequently while the NAND gates are held at their old values the clock transition causes Q to become 1 and $\bar{Q}$ to become 0. If the NAND gates are now allowed to change then Y becomes 1 and Y' remains at 1. Examination of the AND-NOR groups with these new values of Y and Y' shows that no further changes take place in Q or $\bar{Q}$; the circuit is now in a stable condition with $Q = 1$, $\bar{Q} = 0$. A similar analysis for each possible initial

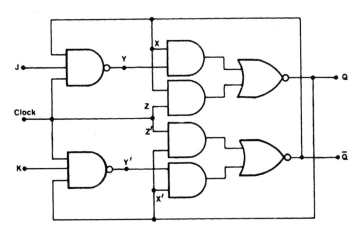

Fig. 4.14 A logic circuit for a negative edge-triggered JK flip-flop

condition at both clock transitions will confirm that the circuit is a negative edge-triggered JK flip-flop provided that the NAND gates operate much more slowly than all the other gates.

A critical examination of all gate delays suggests that the four AND gates should have approximately equal propagation delays. The speeds of the two NOR gates should also be the same. If all the AND and NOR gates have identical delays then analysis of all possible transitions suggests that the NAND gate delays should be at least five times those of other gates.

4.10 Comments

The designer of logic systems to be built with integrated circuits may choose bistables from a large range of readily available devices. Because there are several possible clock control mechanisms the manufacturers' specifications must be fully understood; this may be difficult as such data is sometimes incomplete or is ambiguous.

Usually fewer problems arise when circuits are designed using master-slave flip-flops. However edge-triggered devices must be used for very high speeds or when a particular device is only available in this form. The most important design rule is that all devices in any system should have identical forms of clock control (edge-triggered devices should all be positive edge forms or should all be negative edge forms). This rule may only be broken if all possible timing problems are fully understood and any hazards are detected and eliminated.

Most of the frequently used bistables have been described. One other type which is sometimes met by users of computer aided logic circuit design systems is the T-type (toggle) flip-flop. This has a single control input, T, and a clock input. If $T = 0$ then the Q and $\overline{Q}$ outputs do not change when a clock input occurs, if $T = 1$ then the outputs toggle, that is $Q_{n+1} = \overline{Q}_n$ and $\overline{Q}_{n+1} = Q_n$. The behaviour is exactly the same as that of a JK device with the J and K inputs permanently connected together.

4.11 Problems

1   Use a method similar to that used in Section 4.1 to determine the complete behaviour of the NOR gate SR flip-flop shown in Fig. 4.3.

2   Design the Nth stage of a shift register which can be shifted in either direction. The direction of a shift is determined by the state of a control input.

3   Design the Nth stage of a universal register; i.e. a register which can be shifted in either direction or loaded in parallel. (Hint: two control inputs are required.)

4   Analyse the circuit of Fig. 4.11 by choosing one possible condition at some instant and determining the logic levels throughout the circuit before, during, and after the clock pulse. Repeat for all possible conditions and hence show that the circuit does behave as a JK flip-flop.

5   Figure 4.15 represents a four-stage serial shift register which is constructed using SR negative edge-triggered flip-flops. The Q and $\overline{Q}$ outputs of the last stage are cross-connected to the first-stage inputs as illustrated. If all the stages are initially zero (i.e. Q outputs zero) determine the register contents after each clock pulse input for at least ten clock pulses.

Tabulate your results and comment on the circuit behaviour.

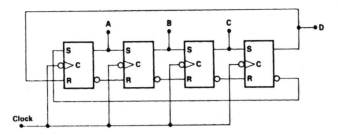

Fig. 4.15

6 Figure 4.16 represents a four-stage serial shift register which is constructed using D-type negative edge-triggered flip-flops. The Q outputs of the last stage and the preceding stage are inputs to an exclusive-OR gate and the output of this gate is the first stage input. Initially, the first stage contains a 1 (i.e. Q is 1) and all other stages are zero; determine the register contents after each clock pulse input for at least sixteen clock pulses.

Tabulate your results. What happens if the initial condition is that all stages *including* the first are zero?

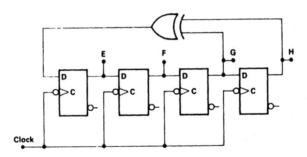

Fig. 4.16

7 Figure 4.13 illustrates a circuit which produces a pulse from a 0 to 1 transition. Devise an alternative circuit which produces a pulse from a 1 to 0 transition.

5 Sequential logic systems

In Chapter 4 sequential logic circuits were classified as either synchronous or asynchronous. When very large circuits are required they are usually designed as an assembly of separate sub-systems; each sub-system is entirely synchronous or entirely asynchronous. The sub-systems are connected so that the overall behaviour of the complete system appears to be totally synchronous or totally asynchronous. Therefore it is always possible to describe any circuit as being either synchronous or asynchronous.

For example most computers have a master clock and are constructed so that all operations take place at times which are precisely determined relative to this clock. Such computers appear to be synchronous systems. Within a computer there are many sub-systems; one of these is an arithmetic unit which performs the calculations and is usually synchronous. Another sub-system is the device by which information (programs, data, etc.) is input to the computer; this is inherently asynchronous because the time at which a user initiates some operation is entirely random with respect to the internal master clock of the computer.

In most cases synchronous circuits are easier to design than asynchronous ones, also fewer hazards are generated by synchronous circuits. Consequently most of the techniques to be introduced for the design of sequential circuits are for synchronous ones (the shift registers already described are synchronous circuits). However it is particularly easy to design asynchronous pure binary counters and their design is examined before more general techniques are described for synchronous circuit design.

5.1 Counters

The term **counter** has a special meaning when used to describe a sequential logic circuit. It refers to a circuit which has a single input, several outputs and behaves in the following manner. Every time a pulse (clock pulse) is input to a counter circuit one or more of the outputs changes. The circuit follows some sequence in which no combination of outputs is repeated until the first one recurs after some fixed number, N, of input pulses – hence there are N different output combinations. When the circuit returns to the original combination of outputs further input pulses cause it to follow exactly the same sequence.

This circuit is called a 'divide-by-N counter' because, if it is arranged (with additional circuits) to produce a single output pulse for every complete cycle of N input pulses, the number of output pulses is the number of input pulses divided by N. Although these circuits are called counters their use is not restricted to counting applications; they form the basis of many sequential logic circuits. For example, each of the different combinations in the N-step sequence may be arranged to initiate some operation in a piece of equipment; N different operations are then performed in a set order. This sequence is called a program and this type of program is typical of those used in simple automatic equipment such as traffic light controllers and automatic clothes washing machines.

Although counters require only one input, many commercial counters have additional inputs called reset (or zero) and preset. The reset input is used to force set the counter to the first position in the sequence regardless of its existing position or input; this is most frequently the condition with all the circuit outputs zero. A preset input sets the counter to some other point in the sequence; usually it is the last one before it returns to the initial zero position.

A **pure binary counter** is one in which the outputs may be regarded as representing a multiple digit binary number; the initial output is the one with all the binary digits (circuit outputs) zero and the counter sequence is such that the binary number increases by 1 for each input pulse until all the outputs are 1 (i.e. after $2^n - 1$ input pulses when the counter has n outputs). The next input pulse returns the counter to zero. Hence 2^n input pulses are required for a complete sequence of an n output (n-stage) pure binary counter. The behaviour of a pure binary counter with three outputs, A, B and C, is given in Table 5.1. Output A represents the least significant binary digit and C represents the most significant digit; as $n = 3$ the counter is a divide-by-eight counter.

Table 5.1

Outputs			Number of input
C (2^2)	B (2^1)	A (2^0)	pulses received
0	0	0	0 (start)
0	0	1	1
0	1	0	2
0	1	1	3
1	0	0	4
1	0	1	5
1	1	0	6
1	1	1	7
0	0	0	8

Binary counters which are *not* classified as pure binary ones are similar to the pure binary type; they are divided-by-N counters which start at zero and follow a binary counting order, but they return to zero after N input pulses where $N \neq 2^m$ (m is any integer).

Non-binary counters do *not* follow a binary counting sequence; they follow some

Table 5.2

Outputs			Number of input
C	B	A	pulses received
0	1	0	0 (start)
1	1	0	1
1	0	0	2
1	0	1	3
0	0	1	4
0	1	1	5
0	1	0	6

other fixed sequence which obeys the rule that no combination of outputs occurs more than once in a cycle of N input pulses and the counter returns to its initial (zero) state after N input pulses. Table 5.2 is an example of one possible divide-by-six non-binary counter; it is one which changes in a Gray code sequence.

The binary counters described so far are such that the number represented by the outputs increases with each input pulse; consequently these are known as **'up' counters**. Alternative designs are possible in which the binary number decreases by one for each input pulse and these are called **'down' counters**. Both types may be combined into a single counter called an **'up–down' counter**; the direction of the count is usually controlled by the logic state at an extra input called a control input. Non-binary counters may also be specified as up or down types if the outputs correspond to the sequence of a well-defined code.

5.2   Pulse sequences

Sequential circuits require some form of pulse input and, although the form of sequence supplied may be complicated, it will usually be one of two types. The sequence may be one in which the pulses occur at intervals which are entirely random in time as in Fig. 5.1a; the pulse width in such a sequence may be random or constant and is usually unimportant. Alternatively the sequence is one which is constant in time and repeats at a fixed interval, it may be a single pulse as in Fig. 5.1b or a more complex scheme as in Fig. 5.1c.

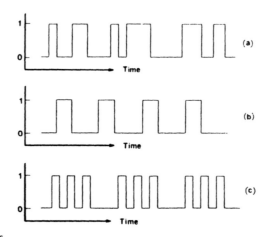

Fig. 5.1   Pulse sequences

The sequence in Fig. 5.1a is similar to the type generated by a detector of random events; e.g. people passing through an entrance, vehicles travelling along a road, or nuclear decay of a radioactive material. If a detector is arranged to indicate such events then a counter circuit connected to the detector output may be used to determine the number of events.

A regular single pulse sequence of the form shown in Fig. 5.1b is usually called a **clock pulse sequence** or clock pulse train. It is frequently used as the input to a counter whose outputs control a sequence of events which occur at precisely determined intervals in time.

5.3 Asynchronous pure binary counters

If the J and K inputs of a JK flip-flop are fixed at 1 it behaves as a divide-by-two counter. Because J = K = 1, each clock pulse input to the flip-flop causes the Q output to change. To produce a 0–1–0 sequence at Q with J = K = 1 requires Q to change twice; i.e. two input clock pulses produce an output sequence of 0–1–0 which is itself a pulse. Thus one output pulse is generated for every two input pulses and the circuit meets the definition of a divide-by-two counter. This is true for both master-slave and edge-triggered types; master-slave devices require the full pulse while edge-triggered devices require one particular edge and ignore the other (which is necessary to restore the clock input level for the next active edge).

A divide-by-N pure binary counter with $N = 2^n$ may be constructed by connecting n master-slave or negative edge-triggered JK flip-flops in cascade. The Q output of one bistable is used as the clock input to the next and all the J and K inputs are fixed at 1. Figure 5.2 shows the logic circuit of a three stage counter constructed in this way. In common with most of the circuit diagrams that are given for examples of sequential logic circuits, the bistable symbols in Fig. 5.2 are those for a negative edge-triggered type. The 'notch' indicates a dynamic input (one affected by a change in logic level, not by the actual level) and the inversion circle indicates that a negative, 1 to 0, change is required. Figure 5.2 also includes the timing diagram for an ideal circuit. Timing diagrams are a useful aid when describing the behaviour of sequential logic circuits, they show the logic states at different points in the circuit on a common time scale.

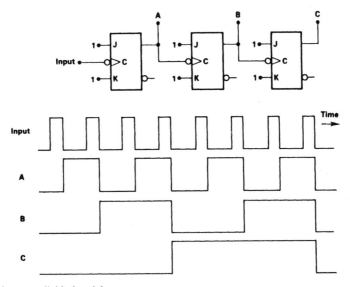

Fig. 5.2 Asynchronous divide-by-eight counter

Figure 5.2 indicates that a single output pulse (0–1–0 sequence) is produced at C when eight pulses are input to the counter circuit. The circuit operation is very simple; every input pulse causes A to change, until A changes there is no clock input signal to B so that any change in B must follow a change in A. Because master-slave or negative edge-triggered flip-flops are used, B only changes when A goes from 1 to 0. Similarly a change in B from 1 to 0 will cause C to change. Thus the action propagates through the counter. Each stage must wait for a change in the previous one before it can begin to change and the circuit operation is clearly asynchronous. This particular design is often called a ripple-through counter or a **ripple counter**. The order in which the outputs of a counter change is shown by a state table; that for a three-stage ripple counter is the same as Table 5.1.

Ripple counters are adequate for many applications and many integrated circuit counters are of this type, but one feature of these counters may cause problems in some applications. The problems arise because each flip-flop takes a finite time to operate (i.e. there is a propagation delay) and the timing diagram in Fig. 5.2 is not strictly correct. If all the flip-flops are assumed to have the same propagation delay t_d and if the output is assumed to change instantaneously after this delay then a corrected timing diagram may be drawn. Figure 5.3 shows part of the corrected diagram.

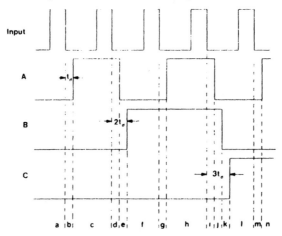

Fig. 5.3   Corrected asynchronous counter timing diagram

Table 5.3

Time	Outputs C	B	A	Decimal value of output	Comments
a	0	0	0	0	Initial state
b	0	0	0	0	
c	0	0	1	1	Correct state but late by t_d
d	0	0	1	1	
e	0	0	0	0	Transient for interval t_d
f	0	1	0	2	Correct state but late by $2t_d$
g	0	1	0	2	
h	0	1	1	3	Corect state but late by t_d
i	0	1	1	3	
j	0	1	0	2	Transient for interval t_d
k	0	0	0	0	Transient for interval t_d
l	1	0	0	4	Correct state but late by $3t_d$
m	1	0	0	4	
n	1	0	1	5	Correct state but late by t_d
		etc.			

With the aid of Fig. 5.3, a table can be constructed which shows the exact circuit behaviour; Table 5.3 is part of the exact table. Examination of this table shows two features of the circuit which were not present when propagation delay was ignored. First, when a correct output state is generated it may occur later than it should by a time of up to $n \times t_d$, where n is the number of flip-flops in the counter. Secondly some output states arise which are not in the correct sequence; they occur only for a brief period which is usually a single propagation delay. Incorrect outputs of short duration are called **transients** or transient outputs and are an additional type of **hazard** situation. Unlike the hazards described previously these do not cause indeterminate or incorrect circuit operation, the circuit eventually reaches the correct state but the incorrect transient outputs may cause problems in large systems.

5.4   An application of an asynchronous counter

Pure binary and other asynchronous counters are frequently used as components in large sequential systems. In electronic systems the transients occur for a very short time and in many applications they do not cause problems. Figure 5.4 shows a typical circuit which incorporates a divide-by-four asynchronous counter; this particular circuit is a controller for a scanner.

The circuit has four outputs, each of which corresponds to one state of the counter. If there is no propagation delay, each output goes to 1 in turn and no two outputs are ever 1 at the same time. A typical use for this circuit is in a monitoring system when readings of four quantities are taken using a single recording instrument which is switched to each signal source in turn. Such a system may be used to record temperatures at different positions in a piece of industrial equipment with a single chart recorder, alarm system or other device. Four temperature sensors (e.g. thermocouples) are switched in turn to the recording device; the switches are controlled by the outputs of the scanner circuit whose clock input is a pulse generator running at a constant frequency which determines the interval between readings.

If the switches used are relays and the scanner driving them is an electronic circuit the transients are unimportant because they will last for less than 0·1 microseconds, whereas

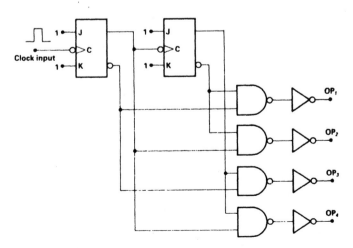

Fig. 5.4   Scanner control circuit

the fastest relays take about 1 millisecond to operate. However, if the switches are themselves electronic devices it is probable that they will operate in a time which is shorter than the duration of a transient output, so the circuit may operate incorrectly.

5.5   Elimination of transient outputs

The most common solution used to remove transient outputs produced in asynchronous circuits is to use a multi-phase clock. For a circuit which is as simple as the scanner in Fig. 5.4, a two-phase clock is adequate. As the scanner is designed for applications where outputs of 0 do nothing, then if all outputs are simultaneously zero there are no problems and clock phase ϕ is used to switch off all the outputs until the circuit has settled into its correct output state. Figure 5.5 shows the circuit for the modified scanner and the timing diagram for the two-phase clock. The main disadvantage of this construction is that the delay, τ, between the two clock pulses must exceed the maximum period for which transients can exist, consequently this type of circuit is inherently slow in operation.

Switching the outputs from an asynchronous circuit so that they appear at a precisely determined time gives the circuit an apparent synchronous behaviour. This technique forms the basis of many methods which are used to connect asynchronous circuits into synchronous systems. Note that there is a small propagation delay in the output NAND gates and inverters. Such delays occur in synchronous circuits and even well-designed synchronous circuits exhibit a slight asynchronous behaviour.

In those cases where the outputs cannot all be switched off until propagation delay effects have disappeared, a two-phase clock solution may still be used. The circuit is built

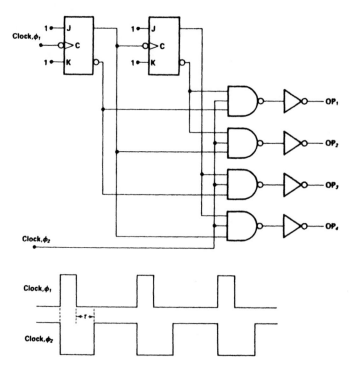

Fig. 5.5   Transient suppressed scanner control circuit

as in Fig. 5.4, but the outputs are used as the inputs to a parallel in-parallel out register, and the final outputs are obtained from the register outputs. Phase ϕ_2 is now used as the clock input to the register and is used to load the register when the circuit has settled in the correct output state.

5.6 Problems

1 If the circuit in Fig. 5.2 is constructed using positive edge-triggered bistables derive a timing diagram and a state table for the circuit in the following cases.
   a) Assuming ideal bistables with no propagation delay.
   b) All bistables have equal, finite propagation delays.
   Design a three-stage asynchronous binary up counter using positive edge-triggered JK bistables.

2 A divide-by-eight asynchronous binary counter is constructed using a particular type of master-slave JK bistable. For this device the delay between the change in the clock input and any consequential change in the Q output is never greater than 20 nanoseconds. If the clock input is a square wave (equal times at 0 and 1) at what clock frequency (approximately) will problems arise such that the counter is of no practical use? Assume that a required (correct) output must exist for at least 20 nsec.

3 A three-stage (divide-by-eight) binary ripple counter is constructed using negative edge-triggered JK flip-flops. These have a set-up time of 5 nsec (minimum) and their propagation delay is 10 nsec when Q changes from 0 to 1 but is 15 nsec when Q changes from 1 to 0. (As J and K are constant hold time is not relevant.) Determine graphically the exact behaviour of the circuit in time when clock pulses 20 nsec wide at intervals of 50 nsec (i.e. clock frequency 20 MHz) are input to the counter. List all the conditions of the counter outputs in the order in which they are produced.

6 The design of sequential logic circuits

Many methods exist for the systematic design of sequential logic circuits; one is described here and is based on a simplified form of Moore's model of sequential circuits. The technique introduced produces a synchronous circuit which is reliable, although it is not always the most economical one possible for a particular application. However, since the design method produces circuits which meet the specification and exhibit very few operating problems it is a useful method for both inexperienced and experienced designers.

6.1 Some definitions

In order to describe any sequential circuit design technique it is useful to define some features of sequential logic circuits in a precise manner.

a) **States or internal states**. A sequential circuit must include one or more internal memory elements; these are bistables (flip-flops) and are denoted by $A_1, A_2, \ldots, A_i$ where i is the number of bistables in the circuit. At any instant in time the circuit may be defined to be in some state (or internal state) denoted by S_α. Each state corresponds to a different set of the Boolean values (logic levels) of the Q outputs of all the bistables. Therefore the system may take any one of the states $S_1, S_2, \ldots, S_j$ where each S_α corresponds to one unique set of values of the outputs of $A_1, A_2, \ldots, A_i$. In general if a system includes i flip-flops it will have 2^i possible internal states, these can be shown in a **state assignment table**. Table 6.1 is one of the ways in which the states could be allocated for a system with two bistables; since $i = 2$ in this case there are four states, S_1, S_2, S_3 and S_4, which may be allocated in any way at all. It is often convenient to allocate the states in order, with the bistable outputs arranged to represent the digits of an increasing binary number as in Table 6.1. However allocation in this way is not essential.

Table 6.1

Q outputs		State
A_2	A_1	allocated
0	0	S_1
0	1	S_2
1	0	S_3
1	1	S_4

In many applications the number of states required is not exactly 2^m (where m is integral). For example a divide-by-five counter requires five states. This is not important; circuits are designed with enough bistables to produce at least the required number of states and, initially, any extra states are neglected. If a system requires j states then the circuit must include at least n flip-flops where $2^n \geqslant j$. In the design method which follows the smallest number of flip-flops possible is used. Therefore n is as small as possible which implies that $2^n \geqslant j > 2^{n-1}$ and hence when j is known n can be determined.

Note that a bistable itself is a sequential logic circuit which has two internal states; the definition here of the state of any sequential circuit is consistent with the description of the state of a bistable made in Chapter 4.

b) **Clock input**. A complete synchronous sequential circuit will usually have a single clock input. Signals at this clock input determine the time at which all changes in the bistable outputs occur; in most cases this is achieved by connecting the circuit's clock input in parallel to the clock inputs of all the bistables in the circuit.

c) **Input conditions**. In addition to the clock input a sequential circuit may have any number of **control inputs**, $B_1, B_2, \ldots, B_k$, although many circuits have no control inputs. The clocked bistables described in Chapter 4 are themselves synchronous circuits with control inputs; for example, the J and K inputs of a JK flip-flop determine (control) its action when the next clock pulse is applied.

The logic levels at the control inputs may be used to define **input conditions**, I_1, $I_2, \ldots, I_i$, where each I_β corresponds to a unique set of Boolean values of the control inputs. Therefore, if there are k control inputs there will be 2^k different input conditions; unlike internal states whose number need not be exactly 2^n, there are always 2^k input conditions.

d) **Circuit outputs**. In most applications the outputs of a sequential logic circuit are used to control the condition of some external system. This control of conditions elsewhere may be achieved directly by the Q outputs of the flip-flops in the sequential circuit, but in many cases additional circuits are needed to produce the outputs required. These output circuits have the flip-flop Q outputs as their inputs and are combinational logic networks for which design techniques have been described previously. The scanner circuit connected to the divide-by-four asynchronous counter described in Section 5.3 is one such output circuit.

Except to note that combinational logic circuits are often required to provide the final outputs of a sequential logic circuit, these output circuits will be ignored at present. Such output circuits may be designed when the design of the sequential section of the network has been completed.

6.2 State diagrams

As in all design problems a complete specification of a sequential logic circuit is required before any attempt can be made to produce a circuit design. A **state diagram** is a useful aid when describing any sequential logic circuit as it is a pictorial representation of the circuit operation.

To construct a state diagram it is necessary to know the number of states and the number of input conditions. Each state is allocated a unique symbol and corresponds to one node in the diagram; it is represented by a circle containing the symbol for that state. From each node there must be a separate flow line for every possible input condition. The flow line has an arrow which shows its direction and is labelled with a symbol to indicate the input condition it represents. The line ends at the node (state) to which the circuit will change when the next clock pulse is input to the circuit. It is essential to show flow lines from every node for every possible input condition, although if two lines are identical, i.e. they are between the same two nodes *and* are in the same direction, they may be combined *but are labelled with both input conditions.* A flow line must be shown even in those cases for which the circuit does not change state, and also for cases which it is considered will never arise; the flow lines leave the node and loop back to it.

To determine the behaviour of some system from its state diagram when a clock pulse is input, all that is necessary is to identify its present state and to follow the flow line for the input conditions which exist. This line will end at the node which corresponds to the state to which the system changes when the next clock pulse input is applied to the circuit.

Example 6.1

Construct state diagrams for two circuits both having three states; S_1, S_2 and S_3, and a single control input which gives rise to two input conditions of I_1 and I_2.

a) The first circuit is a divide-by-three up counter which behaves normally when the control input is 0 and it resets to state S_1 on the next clock pulse if the control input is 1.

b) The other circuit is an up-down divide-by-three counter whose direction is determined by the logic value of the control input.

Solution

In case (a) the input conditions must be assigned. Arbitrarily choose I_1 to be the input condition when the control input is 0, and I_2 to be that when the control input is 1.

In case (b) the choice of control inputs to cause upward and downard counting is left to the designer. Choose a control input of 0 to correspond to upward counting and let this be input condition I_1; the other possible control input of 1 must correspond to downward counting and may be called input condition I_2.

In both cases choose upward counting to be the sequence S_1 to S_2, S_2 to S_3 and S_3 to S_1 (this is just the assignment of states). It is now a simple exercise to draw the two state diagrams; in each case the three symbols for the states are drawn and then flow lines corresponding to I_1 and I_2 are drawn from each state symbol. The completed diagram for case (a) is Fig. 6.1a and that for case (b) is Fig. 6.1b.

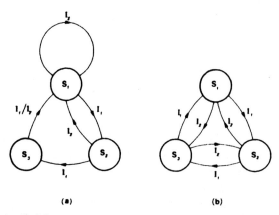

(a)   (b)

Fig. 6.1 Solutions for Example 6.1

Example 6.2

Devise a state diagram for a JK flip-flop. (Since it was stated in Section 6.1 that any bistable is just one particular sequential logic circuit it must be possible to draw a state diagram for any clocked bistable.)

Solution

The flip-flop has two states of $Q = 0$ and $Q = 1$ and has J and K control inputs which will generate four input conditions. The allocation of input conditions and action of the flip-flop for each one are shown in the tables included in Fig. 6.2. The state diagram may be drawn immediately from the information in the table and is also shown in Fig. 6.2. Note how the 'no change' cases are indicated by looped flow lines which leave a state and return to the same state.

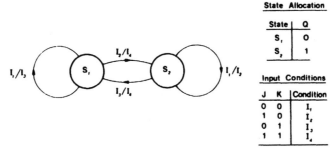

State Allocation

State	Q
S_1	0
S_2	1

Input Conditions

J	K	Condition
0	0	I_1
1	0	I_2
0	1	I_3
1	1	I_4

Fig. 6.2 State diagram for a JK flip-flop

6.3   State tables

The state diagram is a pictorial device which provides an exact and easily interpreted description of the behaviour of a sequential logic circuit. However it is usually difficult to design a circuit directly from a state diagram; a **state table** (transition table) is a more useful aid and may be developed directly from the state diagram for a circuit. State tables may also be constructed without first drawing state diagrams, but there is a much greater possibility of errors being made when there is no state diagram.

In its most simple form a state table consists of a column for the present state, a second column for the input condition, and a third column which shows the state to which the circuit will change when the next clock pulse is received by the circuit. Each row in the state table corresponds to one combination of initial state and input condition. *There must be a row for every possible combination of initial state and input condition*, so that for a circuit with j states and l input conditions there must be $j \times l$ rows in the table. The state diagram of a divide-by-three counter with reset was developed in Example 6.1a; Table 6.2 is the state table for this counter, it has six rows because the counter has three states and two input conditions.

This form of the state table is just a tabular version of the state diagram. A more useful version is one which indicates logic levels. In other words each S_A is replaced by the values of the Q outputs of all of the bistables in the circuit, and each I_B is replaced by the logic levels at all the control inputs. in order to construct this form of state table it is necessary to relate the logic levels at the Q outputs to the states, and those at the control inputs to the input conditions (if these assignments have not already been made). For

Table 6.2

Present state	Input condition	Next state
S_1	I_1	S_2
S_1	I_2	S_1
S_2	I_1	S_3
S_2	I_2	S_1
S_3	I_1	S_1
S_3	I_2	S_1

example the divide-by-three counter with reset in Example 6.1a will incorporate two flip-flops; suppose that their Q outputs are A and B, then choose S_1 to be A = B = 0, S_2 to be A = 1, B = 0, and S_3 to be A = 0, B = 1. If the control input is Z then the original specification in Example 6.1 required that Z = 0 is input condition I_1 and Z = 1 is I_2.

To produce the extended version of the state table the single column for the present state is replaced by a group of columns, these indicate the logic levels at all the Q outputs (one column per flip-flop). The next state column is replaced in a similar manner, and the input condition column is replaced by a group of columns showing the logic levels at all the control inputs. Using the state and input allocations chosen for Example 6.1a its state table, Table 6.2, may be converted into this extended form and becomes Table 6.3.

Table 6.3

Present state		Control input(s)	Next state	
B	A	Z	B	A
0	0	0	0	1
0	0	1	0	0
0	1	0	1	0
0	1	1	0	0
1	0	0	0	0
1	0	1	0	0

This table completely specifies the required circuit behaviour. Although the table appears to be a truth table (and is sometimes called one) it is not one because in a single row of a state table some entries correspond to present logic values of circuit outputs and other entries correspond to future values. A single row in a truth table should only show logic values which exist at the same instant in time.

State diagrams and state tables allow sequential logic circuits to be completely and accurately described; the next step is to introduce a method by which these can be used to produce a design for the circuit.

6.4   Development of an excitation table

Only JK flip-flops may be used as the bistable elements in circuits developed using the detailed design method described here. Additionally in any complete circuit all the JK devices must be the same type; that is all must be master-slave, or all must be positive edge-triggered, or all must be negative edge-triggered. As the circuit is to be a true synchronous one *all* of the flip-flops must have their clock inputs connected directly to a single common source of clock pulses. This ensures that all the flip-flops which are to change on a particular clock pulse do so simultaneously. State tables such as Table 6.3 show how the Q outputs of all the bistables must change when the next clock pulse is input to a circuit which is in a specified state with known input conditions.

For each bistable Q output four different situations may be indicated in the state table. If Q is 0 then it may be required to remain 0 or it may be required to change to 1 when the next clock pulse is input. Similarly if Q is 1 it may remain 1 or change to 0. All four possible cases will be called **transitions**, even though two of the cases involve no change in the Q output. The action of a JK flip-flop for different values of the J and K inputs was summarized in Table 4.5 which is reproduced here for reference.

Table 4.5

J	K	Action at next clock pulse
0	0	No change, $Q_{n+1} = Q_n$
1	0	SET, $Q_{n+1} = 1$
0	1	RESET, $Q_{n+1} = 0$
1	1	Change over (toggle), $Q_{n+1} = \bar{Q}_n$

When the present value of Q is 0 and the state table requires that it remains zero when a clock pulse is input there are two ways in which this can be arranged. If $J = K = 0$ then no change will take place and Q will remain at 0. Alternatively, if $J = 0$ and $K = 1$ a reset will occur and ensures that Q becomes 0 (i.e. remains 0). Hence if Q is 0 and a value of $Q = 0$ is required after the next clock pulse it is necessary to have $J = 0$, but K may be either 0 or 1. This is another type of 'don't care' situation; K may be 0 or $1 - a$ 'don't care' – not because the conditions will never arise, but because with either value at the K input the bistable will operate as required by the state table.

Thus to obtain the output transition of 'Q is 0 and becomes 0', the control inputs required by a JK flip-flop are $J = 0$ and $K = $ 'don't care' $= X$. A similar argument for the transition 'Q is 0 and becomes 1' gives the requirement that the control inputs are $J = 1$ and $K = X$. Examination of all four possible output transitions to determine the values of J and K required to produce each one leads to the results summarized in Table 6.4.

Table 6.4

Present output Q_n	Next output Q_{n+1}	Control required J	K
0	0	0	X
0	1	1	X
1	0	X	1
1	1	X	0

Table 6.4 is the **excitation table** for a JK flip-flop and contains all the information required to extend the state table of a circuit so that it becomes an excitation or switching table. An excitation table for a sequential logic circuit constructed using JK flip-flops shows the logic levels required at every J and K input to produce the correct Q outputs

when the next clock pulse is input. If J_A and K_A are the control inputs of the flip-flop whose Q output is called A in Table 6.3 then, using Table 6.4, the values of J_A and K_A needed to produce the required change in A may be determined for each row in the truth table. For example, in the first row of Table 6.3 output A changes from 0 to 1; examination of Table 6.4 indicates that to produce this transition J_A must be 1 and K_A is a 'don't care'. In the same row of the table B is required to make the 0 to 0 transition which requires $J_B = 0$ and $K_B = $ 'don't care' = X.

By using this technique of examining how each flip-flop Q output is required to change in every row in a state table, the corresponding excitation table can be produced. This is a state table to which further columns have been added; these columns show the values required at all the J and K inputs to produce the specified changes in the Q outputs. In the case of Example 6.1a, for which Table 6.3 is the state table, this determination of J and K inputs produces Table 6.5 as the excitation table.

One step remains to complete this table. The example requires two bistables, hence four states are possible, but only three are used. To complete the circuit design it is necessary to specify values for the J and K inputs corresponding to all possible combinations of initial state and input condition, including the impossible condition of the circuit being in the unused or redundant state. (Other systems may have several unused states or none.) At present it is assumed that the circuit can never be in this unused state so that it does not matter what values J and K take; i.e. both J and K are 'don't cares'. The two additional rows shown in Table 6.6 must be added to Table 6.5.

This addition of 'don't cares' for unused states leads to some problems in a very small number of cases; these problems are examined in Section 6.7.

Table 6.5

Present	state	Control	Next state		Flip-flop controls			
B	A	Z	B	A	J_B	K_B	J_A	K_A
0	0	0	0	1	0	X	1	X
0	0	1	0	0	0	X	0	X
0	1	0	1	0	1	X	X	1
0	1	1	0	0	0	X	X	1
1	0	0	0	0	X	1	0	X
1	0	1	0	0	X	1	0	X

Table 6.6

Present	State	Control	Next State		Flip-flop controls			
B	A	Z	B	A	J_B	K_B	J_A	K_A
1	1	0	X	X	X	X	X	X
1	1	1	X	X	X	X	X	X

6.5 Design of the circuit

An excitation table contains all the information required to complete the design of a sequential logic circuit. As stated previously all the bistable clock inputs must be connected to the same source of clock pulses, this source controls the circuit timing. All that remains is to design circuits which will provide the required J and K inputs for each flip-flop.

All the JK flip-flops in the circuit must be of the same type. As their clock inputs are connected to a common clock source then when several Q outputs change as the circuit moves from one state to the next they all change simultaneously. These changes are defined by the values of the J and K inputs which must be constant throughout the 0–1–0 clock pulse if master-slave bistables are used; alternatively J and K are fixed through the set-up, transition, and hold times of the active clock transition for edge-triggered devices. Because all changes occur simultaneously the existing (present state) Q outputs may be used together with any external control inputs to define J and K, (i.e. J and K derived from the present Q values will remain constant until all the new Q values appear; when these new values appear the J and K values used to generate them are no longer required). In this situation the J and K values may be provided from the outputs of combinational circuits whose inputs are the present Q outputs and the external control inputs. Strictly a check should be made that the specification for the bistables indicates that their maximum hold time is less than their minimum propagation delay. In general this requirement is met as integrated circuit manufacturers design bistables for use in synchronous sequential circuits.

If different types of JK flip-flops are used within one circuit their outputs will change at different times; as the outputs of one bistable may control the J or K inputs of other bistables which require the previous (unchanged) value incorrect operation might occur. It is possible to develop design techniques allowing a mixture of bistable types to be used in a circuit. However such techniques would be difficult to use, not easily put into a general form, and design errors could lead to race hazards which are not obvious. The technique described here requires that a single type of JK bistable is used throughout any circuit; mistakes in design produce circuits with obvious incorrect operation. The single type of JK bistable may be of any form, circuit diagrams have been drawn assuming that negative edge-triggered devices are used.

The truth tables for the combinational logic circuits are contained within the excitation table. For example, the first row of Table 6.5 shows that when $A = B = Z = 0$ the four combinational logic circuits which are required must have outputs which generate the required inputs of $J_A = 1$, $K_A = X$, $J_B = 0$ and $K_B = X$. Similarly the second row indicates that when $A = B = 0$ and $Z = 1$ these combinational logic circuits must give outputs which supply $J_A = 0$, $K_A = X$, $J_B = 0$ and $K_B = X$. Therefore the excitation table can be used to provide truth tables for combinational logic circuits. There must be one truth table corresponding to the circuit which supplies a single J input, and another for the circuit supplying one K input; i.e. there are two truth tables and two circuits for every bistable. Usually the excitation table is used directly to give Boolean expressions for the combinational circuits, but to illustrate the procedure fully the truth table for the circuit whose output is connected to the J_A input in the example is given in full as Table 6.7.

The combinational logic circuit which is required to provide the correct J_A input must obey this truth table; it may be designed using the techniques developed in Chapter 3.

Table 6.7

| Present state | | Control | |
B	A	Z	J_A
0	0	0	1
0	0	1	0
0	1	0	X
0	1	1	X
1	0	0	0
1	0	1	0
1	1	0	X
1	1	1	X

The most suitable technique in most cases is to draw the Karnaugh map and then use it to deduce a minimal Boolean expression for the circuit. Maps for all four circuits required by excitation Table 6.5 are shown in Fig. 6.3. The groups selected on the maps require

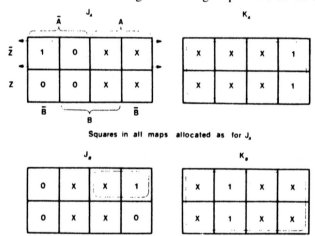

Squares in all maps allocated as for J_A

Fig. 6.3

$J_A = \bar{B} . \bar{Z}$, $K_A = K_B = 1$ and $J_B = A . \bar{Z}$. Thus the circuits for the J and K control inputs to the flip-flops have been determined. The complete circuit diagram may now be drawn (Fig. 6.4); note the use of the flip-flop $\bar{Q}$ outputs to provide inverted quantities.

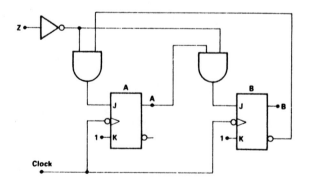

Fig. 6.4

6.6   Summary of the design method

The design technique described may be applied to any synchronous sequential logic circuit and the following list is a step by step summary of the method.

a)   Formulate the problem clearly; that is produce a complete specification for the circuit.

b)   Decide how many states and how many control inputs are required.

c)   Determine the minimum number, n, of bistables required from the relationship $2^n \geqslant j > 2^{n-1}$, where j is the number of states.

d)   Allocate states to the output conditions of the flip-flops (i.e. devise a state allocation table). This allocation is arbitrary, but it is often convenient – and frequently produces economical circuits – to allocate the states so that the flip-flop Q outputs represent a multiple digit binary number which increases by one for each step in the most common sequence followed by the circuit.

e)   Allocate input conditions to all possible combinations of logic values at the control inputs.

f)   Draw an exact state diagram for the circuit.

g)   From this state diagram form a complete state table for the circuit; add 'don't care' conditions in the next state columns corresponding to unused initial states.

h)   Convert the state table into an excitation table so that the J and K inputs to all the flip-flops are specified in all cases.

i)   Determine the combinational logic circuits required to produce these values of J and K as outputs; the inputs to these circuits are all the flip-flop Q outputs and all the control inputs.

j)   Draw the complete circuit diagram.

k)   Construct and test a prototype circuit.

Example 6.3
   Design a divide-by-six up counter.

Solution
   A detailed solution is given to this problem.

a)   The problem clearly specifies the circuit.

b)   A divide-by-six counter must have six states and as it is a single-direction counter with no special features it has no control inputs.

c)   There are six states hence the number of bistables, n, is given by $2^n \geqslant 6 > 2^{n-1}$ so that n must be three.

d)   Table 6.8 shows the state allocation selected; A, B and C are the Q outputs of the three bistables.

Table 6.8

State	C	B	A
S_1	0	0	0
S_2	0	0	1
S_3	0	1	0
S_4	0	1	1
S_5	1	0	0
S_6	1	0	1

e) As there are no control inputs there are no input conditions to allocate.

f) The state diagram is a simple one and is shown in Fig. 6.5.

g) and (h) From Fig. 6.5 the state table is constructed and the columns required to convert it to an excitation table are added; the result is Table 6.9.

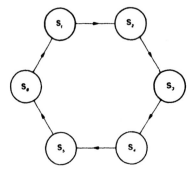

Fig. 6.5   A state diagram for a divide-by-six counter

i) Using Table 6.9 the Karnaugh maps for each J and K input circuit are completed. These maps are shown in Fig. 6.6 and the solutions found are $J_A = K_A = 1$, $J_B = A.\bar{C}$, $K_B = A$, $J_C = A.B$ and $K_C = A$.

j) Using these relationships the circuit diagram is drawn and is included in Fig. 6.6.

Table 6.9

Comment	Present state			Next state			Flip-flop controls					
	C	B	A	C	B	A	J_C	K_C	J_B	K_B	J_A	K_A
$S_1 \to S_2$	0	0	0	0	0	1	0	X	0	X	1	X
$S_2 \to S_3$	0	0	1	0	1	0	0	X	1	X	X	1
$S_3 \to S_4$	0	1	0	0	1	1	0	X	X	0	1	X
$S_4 \to S_5$	0	1	1	1	0	0	1	X	X	1	X	1
$S_5 \to S_6$	1	0	0	1	0	1	X	0	0	X	1	X
$S_6 \to S_1$	1	0	1	0	0	0	X	1	0	X	X	1
Unused	1	1	0	X	X	X	X	X	X	X	X	X
Unused	1	1	1	X	X	X	X	X	X	X	X	X

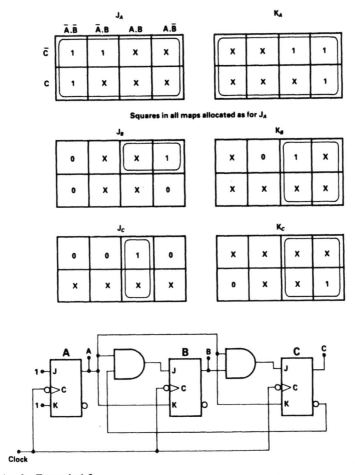

Fig. 6.6 Solution for Example 6.3

Example 6.4

Design a divide-by-four up–down counter.

Solution

Only a brief outline of the solution is given for this example.

The circuit behaviour is clearly defined in the problem and it is apparent that two bistables are needed to produce the four states required. The method of selecting the count direction has not been specified and may be freely chosen; an obvious choice is to use a single control input to produce two input conditions. Both the state and input condition allocations chosen are given in Table 6.10 while Fig. 6.7 and Table 6.11 show the state diagram and corresponding excitation table respectively. The Karnaugh maps for the circuits which supply the J and K inputs and the final circuit diagram are given in Fig. 6.8.

Table 6.10

State	B	A
S_1	0	0
S_2	0	1
S_3	1	0
S_4	1	1

(a) State allocation

Control input	Input condition	Comment
0	I_1	Count up
1	I_2	Count down

(b) Input conditions

Table 6.11

Present state B	A	Control Z	Next state B	A	J_B	K_B	K_A	K_A
0	0	0	0	1	0	X	1	X
0	1	0	1	0	1	X	X	1
1	0	0	1	1	X	0	1	X
1	1	0	0	0	X	1	X	1
0	0	1	1	1	1	X	1	X
0	1	1	0	0	0	X	X	1
1	0	1	0	1	X	1	1	X
1	1	1	1	0	X	0	X	1

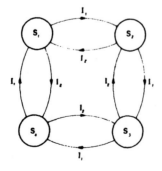

Fig. 6.7

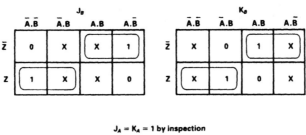

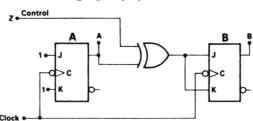

$J_A = K_A = 1$ by inspection

Fig. 6.8

6.7 Problems associated with unused states

It was indicated in Section 6.4 that operating problems may arise in circuits which have been designed with 'don't care' conditions in the next state column of the state table for cases when the initial state is an unused (redundant) one. The problems arise because at switch on each bistable settles at random with either possible value of Q output and the circuit takes up any possible state including the unused ones.

Often there is no problem as a few clock pulses drive the circuit from any unused state into a used one and thereafter the circuit follows the correct sequence. However in some cases the circuit may become trapped in an unused state or group of unused states. Three different state diagrams for a divide-by-five up counter with three unused states S_6, S_7 and S_8 are shown in Fig. 6.9; the unused states are shown in three of the many possible ways.

In case (a) the circuit cannot become trapped in an unused state but in the other two cases it may be trapped.

Several methods are available to avoid or overcome this particular problem.

a) After power is switched on a manual reset switch is used; this is connected so that the circuit is forced into a known used state. This method, or the next one, is commonly used with very large systems and has the additional advantage that the circuit starts at a known position in its sequence.

b) Method (a) is automated; an electronic or electromechanical device senses the rise in supply voltage at switch on and provides the forced reset when the voltage reaches the correct operating level.

c) The problem is ignored and the circuit is designed using 'don't cares' for the next states corresponding to unused initial states as described in Section 6.6. A state diagram which includes the unused states is then drawn and indicates the behaviour of the circuit when in any unused state. Possible problems are easily identified from this diagram; the diagram for the circuit designed in Example 6.3 is shown in Fig. 6.10 and indicates that the circuit will not become trapped.

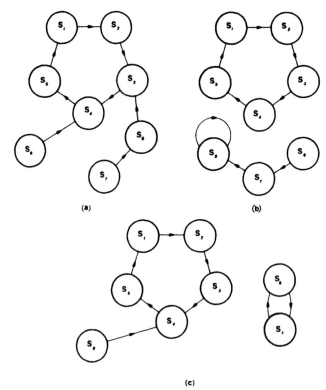

Fig. 6.9   Some state diagrams for a divide-by-five counter

If problems are found (a surprisingly rare occurrence) it is necessary to use solutions (a) or (b) or to start the design again using the following method.

d)   This technique may be used initially instead of the method summarized in Section 6.6, or may be used if that method produces a circuit which can become trapped in an unused state. The technique is to draw the initial state diagram with the unused states included; flow lines are shown leaving these unused states for all possible input conditions. No flow line must end at an unused state. This state diagram is used as the basis of the state and excitation tables which are prepared and then used as in Section 6.6; the circuit produced is one which cannot become trapped in an unused state.

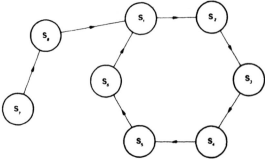

Fig. 6.10   State diagram, including unused states, for the result of Example 6.3

The disadvantage of the technique is that the circuit will usually be less economical than one designed by the method which initially ignores unused states. It is probable that more gates will be required to construct the combinational circuits which generate the J and K inputs.

Example 6.5

Design a synchronous divide-by-six up counter arranged so that any transitions from unused states end at the first used state.

Solution

This is Example 6.3 repeated with the modification that unused states are not neglected. Steps (a) to (e) in the design process are identical but at step (f) a revised state diagram is required and is shown in Fig. 6.11. The new excitation table is easily derived; it is Table 6.12 and the expressions derived from it for the connections to the J and K inputs are $J_A = \bar{B} + \bar{C}$, $K_A = 1$, $J_B = A.\bar{C}$, $K_B = A + C$, $J_C = A.B$ and $K_C = A + B$. Some of these expressions are obviously more complex than the corresponding ones obtained in Example 6.3. However any circuit constructed using these expressions will not stick in an unused state; its behaviour is completely specified and no further checks are required.

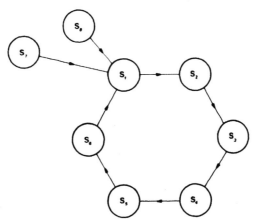

Fig. 6.11 State diagram specified in Example 6.5

Table 6.12

Present state			Next state			Flip-flop controls					
C	B	A	C	B	A	J_C	K_C	J_B	K_B	J_A	K_A
0	0	0	0	0	1	0	X	0	X	1	X
0	0	1	0	1	0	0	X	1	X	X	1
0	1	0	0	1	1	0	X	X	0	1	X
0	1	1	1	0	0	1	X	X	1	X	1
1	0	0	1	0	1	X	0	0	X	1	X
1	0	1	0	0	0	X	1	0	X	X	1
1	1	0	0	0	0	X	1	X	1	0	X
1	1	1	0	0	0	X	1	X	1	X	1

6.8   Design from timing diagrams

So far the sequential circuit designs examined have been produced to meet a written specification. When a large system is designed by a team an individual designer may be required to produce a sub-system for which the specification is a timing diagram. This is because in a large system there is usually a master clock (often multiphase) which produces a continuous clock pulse train that controls the time at which all changes occur.

The range of timing diagrams for possible circuits is infinite; the following example illustrates one approach to a particular circuit design.

Example 6.6

Figure 6.12 shows the master clock of some system and the two output pulse sequences required from the single output of a circuit. Design a circuit so that the output sequence produced depends upon the condition at a single control input Z.

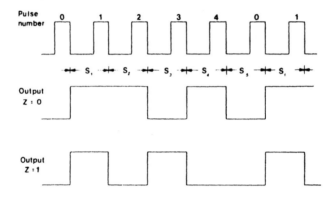

Fig. 6.12

Solution

This is a simple design problem as both output sequences in Fig. 6.12 have the same cycle length (repetition interval). This cycle length can be determined in terms of a number of clock pulses – five in this example – and this number is also the number of states (N) required in the sequential circuit. Once the cycle length is known a divide-by-N single direction counter is designed using any standard technique. In cases for which variable length sequences are required the counter must have control inputs which are connected so that the cycle length depends upon the input conditions. In the example output changes occur on the 1 to 0 input change, therefore in this case the counter must be constructed with master-slave or negative edge-triggered bistables.

The final output is produced by a combinational logic circuit which has the control input (or inputs) and the bistable Q outputs as its inputs. This is the type of output circuit which was briefly introduced in Section 6.1d. For the example let states S_1, S_2, S_3, S_4 and S_5 correspond to the intervals which occur after clock pulses 0, 1, 2, 3 and 4 respectively. The output circuit truth table is constructed and shows the final circuit output required from the combinational circuit for all possible combinations of input conditions and bistable Q outputs. Unused states are considered as never existing and the final output corresponding to such states may be taken as 'don't care'.

Table 6.13

| State | Counter flip-flop outputs | | | Control | Circuit |
	C	B	A	Z	Output
S_1	0	0	0	0	1
S_1	0	0	0	1	1
S_2	0	0	1	0	1
S_2	0	0	1	1	0
S_3	0	1	0	0	0
S_3	0	1	0	1	1
S_4	0	1	1	0	1
S_4	0	1	1	1	0
S_5	1	0	0	0	0
S_5	1	0	0	1	0
Unused	1	0	1	0	X
Unused	1	0	1	1	X
Unused	1	1	0	0	X
Unused	1	1	0	1	X
Unused	1	1	1	0	X
Unused	1	1	1	1	X

Table 6.13 is the output circuit truth table for the example and may be used to produce a minimal logic expression for the circuit output. The result obtained is

$$R = A.\bar{Z} + \bar{A}.\bar{B}.\bar{C} + \bar{A}.\bar{C}.Z = \overline{(A.\bar{Z}).(\overline{\bar{A}.\bar{B}.\bar{C}}).(\overline{\bar{A}.\bar{C}.Z})}$$

When a combinational circuit which implements this is connected to the divide-by-five counter the circuit in Fig. 6.13 is obtained (details of the counter itself are not shown).

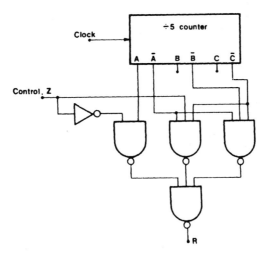

Fig. 6.13 Solution for Example 6.6 (omitting details of counter)

6.9   Comments

This treatment of sequential circuit design provides a relatively simple technique; the main disadvantage is that at no stage has the restriction been imposed that the most economical circuit should be produced. Some more advanced techniques attempt to minimize the circuit but can rarely take into account all possible reductions.

For example, a simple divide-by-four counter requires four states and the method described here allows the states to be allocated to the values of the bistable Q outputs in an entirely arbitrary manner. It was suggested that a systematic allocation be used but this is not essential; in a simple four-state circuit there are twenty-four different ways in which the allocation could be made. Any more advanced technique should determine the allocation which leads to the most economical solution. However as electronic components are inexpensive and very reliable this non-minimal solution is not a major disadvantage when the technique described is used for electronic systems. It is more important to produce a hazard-free circuit than to impose any other limitation on the design of a sequential circuit.

The design technique may be adapted to utilize other types of bistable. All that is necessary is to derive the excitation table (equivalent to Table 6.4 obtained for the JK flip-flop) for the chosen flip-flop then modify the method accordingly. Generally circuits designed using SR or D-type flip-flops will require more elaborate combinational circuits to generate the flip-flop control inputs than would be required for circuits using JK flip-flops.

6.10   Problems

1   Produce a circuit diagram for a divide-by-five counter (single direction only) constructed using JK master-slave flip-flops.

2   Design a two-directional (up–down) divide-by-six counter constructed from JK positive edge-triggered flip-flops. Determine the behaviour of your circuit if it gets into any unused state.

3   Design a single-direction counter using three JK flip-flops. The counter must have five states (i.e. it must be divide-by-five up counter) but the counting sequence must be such that if the three Q outputs are taken as the digits of a BCD number, the count sequence is 0, 1, 3, 5, 6, 0, 1, 3 etc.

4   Derive the circuit diagram of a synchronous counter with a single control input; when the control input is 1 the counter behaves as a divide-by-seven up counter and when the control input is 0 the behaviour is a divide-by-five up counter.

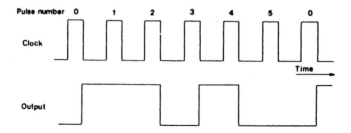

Fig. 6.14

5 Figure 6.14 is a timing diagram showing the clock input and single output of a sequential logic circuit. Devise a circuit which behaves as required by this diagram.

6 Devise a sequential logic circuit to control traffic lights (traffic signals) at a simple crossroads junction. Assume that a clock pulse is provided every two seconds and that while changing, the lights remain at an intermediate state for two seconds. The lights operate on a simple time control sequence (no traffic-actuated control); traffic on each road is allowed to move in turn and the lights remain green for ten seconds.

7 Design a divide-by-four up–down counter using clocked SR flip-flops.

8 Design a single direction counter such that the Q outputs of three bistables when regarded as a binary number follow the sequence 0,1,2,3,5,7,5,3,2,1, which then repeats. (Hint: the circuit probably requires more than eight states.)

9 The controller for an industrial process is based on a sequential logic circuit; the state of the circuit is used to decide the operations of a safety valve and a heater. The control inputs to the circuit are from two switches indicating pressure too high and temperature too low.

All possible conditions of control input may exist and the system may be in any one of the four states comprising all combinations of safety valve open or closed, and heater on or off.

Devise a state diagram for the sequential circuit and design a circuit which behaves as required by this state diagram. Also design the output combinational circuit which uses all the bistable Q outputs to produce signal control for the safety valve and heater.

7 Electronic logic circuits

In the preceding chapters it was usually assumed that logic circuits could be constructed using ideal components whose internal construction imposed no limitations on circuit design. Under such conditions the only restrictions which apply to the design of logic networks are those set by the laws of Boolean arithmetic and algebra. Any real system will be further restricted because it must be built with components with a performance which is not ideal and designers must make allowances for component limitations.

When studying logic circuits it is essential to perform exercises involving their design, construction and testing. Generally either 7400 series TTL or 4000 series CMOS integrated circuits are used for such exercises; comments concerning the properties and limitations of electronic components apply only to these two series of devices. However, other forms of electronic logic device, and most non-electronic ones, will have features that impose similar constraints on circuit design and operation. The reasons for the non-ideal behaviour of electronic logic elements requires detailed examination of the construction of integrated circuits. This is usually a separate major topic within engineering courses and is not examined here.

Features of 7400 and 4000 series devices are described when required; the reader should refer to manufacturers' data for full information and for numerical values of limiting parameters for individual devices. A partial explanation of manufacturers' code numbers is included later in Table 7.2.

It is convenient to divide the problems which arise because components do not have ideal behaviour into two groups. There are problems because elements require a finite time to operate and there are interconnection problems. One cause of interconnection problems is that the output of any electronic circuit can only provide input signals to a limited number of further circuits. (Any signal transmission requires transfer of energy and an output can supply only a finite amount of energy.) This, and other interconnection problems, leads to the introduction of special components which are not strictly two-state logic devices. These components allow some problems to be solved more economically than is possible using true two-state components. Common special devices include those with open collector outputs (also open drain) and those with tri-state outputs.

7.1 Timing problems

Some circuit problems arise because real logic elements take a finite time to operate (**propagation delay**) and one or two of these problems have already been described. Timing hazards are an inherent feature of all practical logic systems and two hazards have already been examined. These are the race condition (or critical race) described in Section 4.1 and the production of transient outputs by asynchronous circuits examined in Section 5.3.

7.1.1   Race conditions

Strictly speaking race conditions do not arise because logic elements exhibit propagation delay. A problem in predicting circuit operation exists whenever two signals simultaneously attempt to produce opposite effects and this problem would be present even if the elements had zero propagation delay. The effect of finite, but marginally different, propagation delays often allows the sysem to settle into an unknown but stable state as in the case of the SR flip-flop with S = R = 1 examined in Section 4.1. In other cases the circuit may oscillate and Fig. 7.1 shows an example of one such circuit; it consists of five inverters connected in a single closed loop and a simple investigation illustrates the race condition. Assume that a particular logic level is present at the input to one gate and determine the conditions at each output around the loop; the analysis indicates that the opposite level to the one assumed should be present at the original input. Because the elements have a finite propagation delay the circuit oscillates with a period of oscillation which is approximately ten times the mean propagation delay for a single gate.

If this explanation is found to be difficult replace one inverter by a two input NAND gate. One NAND input just completes the loop, as with the inverter, the other is connected to a switch which can supply 0 or 1 to it. Assume that the input from the switch has been at 0 for a long time and is then changed to 1 and remains at 1. Determine the circuit behaviour at the NAND gate output as a function of time using the method introduced later in Section 7.1.2.

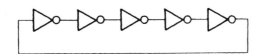

Fig. 7.1   Five-inverter loop oscillator

If perfect (i.e. zero propagation delay) elements were used to construct an SR flip-flop and a five-inverter loop the exact circuit behaviour for the race conditions could only be determined with knowledge of the internal construction of the elements. It is possible that in such circumstances the circuit would settle into a condition with the outputs outside the range defined for the logic levels.

The race conditions are obvious in these two simple examples but it is often difficult to identify possible race hazards in large networks. While circuits designed correctly using the techniques developed in Chapter 6 should not contain race hazards, the methods are not easily applied to very large networks. Often large networks are divided into smaller sections, each section is designed separately, and then the separate designs are interconnected. Race hazards may be introduced and are not easily detected; the most successful method of identifying these, and other hazards, is to program a computer to simulate the network operation under a wide range of conditions. Programs of this form have been devised and their output is such that possible problems are indicated.

7.1.2   Transient outputs

The production of transient outputs by asynchronous circuits is a major problem associated with such circuits. Transients of very short duration may even arise in well-designed synchronous circuits if the circuit components have widely differing propagation delays. However transient problems with synchronous circuits are uncommon if all the components within the network are chosen from a single logic family.

Transient outputs may also be produced by combinational networks. The circuit of Fig. 7.2 illustrates this; it implements the function

$$R = A.\bar{B} + \bar{A}.C$$

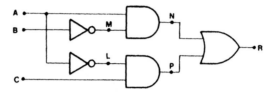

Fig. 7.2

Suppose that at some instant in time, t_0, the circuit input conditions change instantaneously from $A = 1$, $B = 0$, $C = 1$ to $A = 0$, $B = 0$, $C = 1$; i.e. A changes. Assume that all the logic elements have an identical propagation delay of t_d then at times between t_0 and $t_0 + t_d$ input A will have changed but the output, R, and all the intermediate points marked L, M, N and P will remain unchanged. At the time $t_0 + t_d$ those gates which are connected directly to the input A will have changed but those connected indirectly (i.e. one or more gates between A and the element) will not have changed. This examination can be repeated at successive intervals of t_d until there are no further changes in any gate outputs. Full details of this analysis are shown in Table 7.1; entries marked with an asterisk indicate cases for which a gate output has not had time to change and the output is not the one implied by the Boolean function for the gate with the input conditions shown.

Table 7.1

Time, t, from $t_0 = 0$	A	B	C	$L = \bar{A}$	$M = \bar{B}$	$N = A.M$	$P = L.C$	$R = N+P$	Comments
Before t_0	1	0	1	0	1	1	0	1	Initial conditions
$t_d > t > 0$	0	0	1	0^*	1	1^*	0	1	New input, L and N must change
$2t_d > t > t_d$	0	0	1	1	1	0	0^*	1^*	L and N have changed, this requires that P and R change.
$3t_d > t > 2t_d$	0	0	1	1	1	0	1	0^*	P and R have changed but the new value of P requires R to change again.
After $t = 3t_d$	0	0	1	1	1	0	1	1	Final condition

Examination of the column for the output R in Table 7.1 indicates that although both the initial and final input conditions give an output of 1, a different output of 0 is produced for an interval of t_d. Thus a transient output is produced by a combinational logic circuit; such transients and more complex ones (e.g. a sequence $0 - 1 - 0 - 1$ for a simple 0 to 1 change) are quite common in very large combinational networks. Removal of the restriction that all gates have identical propagation delays complicates the analysis but does not significantly alter the results.

Design techniques exist which produce 'transient free' combinational circuits but only work if circuits are constructed using gates selected with approximately the same propagation delays. Such 'transient free' circuits have the additional disadvantage that they are often less economical than circuits designed ignoring the possible production of transients.

Transient outputs may cause difficulties when circuits which produce them are used to supply inputs to sequential circuits. The most convenient method of overcoming such difficulties is to use a two phase clock as illustrated by the asynchronous counter described in Chapter 5. In the most simple arrangement the new inputs are connected at a time set by clock phase ϕ_1, the circuit outputs are connected to further circuits at a later time set by clock phase ϕ_2. The interval between ϕ_1 and ϕ_2 is set to be greater than the maximum interval in which transients may occur. In complex systems it may be necessary to use a multi-phase clock rather than a two-phase one.

7.2 Interconnections

One assumption made previously is that the output of one logic element is capable of being connected to supply the inputs to any number of other elements. In any real system (electronic or other) the output of one element can only drive a finite number of inputs of further elements; this finite number is termed the **fan-out** of the element. It is possible to make this definition because the standard elements in any family of logic devices are designed so that all inputs are the same (require the same electrical signals).

Not all manufacturers define the ability of one output to drive further inputs in terms of fan-out; even those who use fan-out do not use identical definitions and care is necessary when using manufacturers' data. The method adopted here is to define a standard unit called a **unit load** (UL); the fan-out capability of any element is measured in unit loads. The value selected as one unit load is the input requirement of a standard device of the original 7400 series. Generally all the components in a series have the same fan-out; however most series include some devices called **buffers** which have a much higher fan-out and are used to simplify construction when one output is required to drive a very large number of inputs.

As integrated circuit logic families are developed some components are produced with inputs that do not have the standard drive requirement of the series. Therefore input requirements are also measured in terms of the unit adopted for fan-out; the requirement of any input is called its **fan-in**. For example many JK bistables have J and K inputs with the standard fan-in for their series but the clock input has a fan-in which is twice this. When these devices are used the output of a gate can drive twice as many J and K inputs connected in parallel as it can drive clock inputs connected in parallel.

The original basic 7400 series should be regarded as obsolete although components are still available for repair purposes. Even the newer 74LS00 series is generally regarded as too old for use in new designs intended for manufacture. A bewildering range of 74x00 series exists; the letter x is replaced by a one-, two- or three-letter code which identifies the series. In pure logic terms all the series are identical, the logic function of any device is defined by the code number used in place of 00 (code 00 itself defines a device with four two-input NAND gates in a single integrated circuit package). At the time of writing the 74HCT00 series is suggested as a reasonable choice for study exercises and general manufacturing applications (however do not use the 74HC00 and 74AC00 series as they are not compatible with other 74x00 series). Table 7.2 outlines the properties of some of the more common 74x00 series and includes a partial explanation of the device code numbers. Note that the concept of fan-in does not apply for the 74HCT00 and 74ACT00 series as their inputs require very little drive current. Theoretically an output of one unit load can drive several hundred 74HCT00 series inputs in parallel; however a fan-in value in the region of 0·025 to 0·1 of a unit load should be regarded as a practical limit for 74HCT00 series devices.

In general all elements in a logic network should be from a single series. In this case interconnection between logic elements is straightforward and it is only necessary to check that the total of the fan-ins of all the inputs connected to each output does not exceed the fan-out of the output. For most modern 74x00 series, except the 74HCT00 series, the ratio of fan-out to fan-in is twenty to one; however manufacturers' data should be checked for any non-standard fan-in and fan-out ratings of individual devices.

Although all devices in a logic network should be from a single series a few logic functions are only manufactured in one series. When these functions are required it may be necessary to mix devices from different series (these must be 74x00 series and must not include the 74HC00 or 74AC00 series). In such cases detailed manufacturers' data must be examined carefully, the connections can be made if the fan-out of the driving device is greater than the total of the fan-ins of all devices to which it is connected.

Example 7.1

Determine if each of the following connections will operate reliably assuming that all components have the normal fan-in and fan-out ratings for their series.

a)   A 74LS output driving three 74AS inputs and one 74ALS input.

b)   A 74HCT output driving four 74F, five 74HCT and two 74AS inputs.

c)   A 74F output driving two 74LS and three 74ALS inputs.

Solution

a)   From Table 7.2 the 74LS output has a fan-out of 5·0 unit loads (ULs). Each 74AS input is 1·25 ULs so the three form 3·75 ULs. The 74ALS input is 0·25 UL so the total drive requirement is 4·0 ULs. This is within the 74LS fan-out rating and the connection may be made.

b)   As for part (a) compute the total load which is $4 \times 0·5 + 5 \times 0·025 + 2 \times 1·25$, that is 4·625 ULs. This exceeds the 74HCT output capability and the connection must not be made.

c)   The total load is 1·25 ULs, this does not exceed the capability of the 74F series device and the connection may be made.

Interconnection rules for 4000 series CMOS devices are not quite so simple as those for 7400 series TTL devices. In theory a CMOS output is capable of driving an infinite number of further CMOS inputs; in practice there is a finite, although large, limit. This limit cannot be described by a single number but must be determined for each application because the fan-out is determined by two operating conditions selected by the circuit designer. These conditions are the supply voltage (not fixed for CMOS unlike TTL) and the propagation delay; their relationship to fan-out is illustrated by Fig. 7.3.

Hence, to obtain a numerical value for the fan-out of a CMOS element the supply voltage and maximum tolerable propagation delay must be specified. The fan-out is then

Table 7.2

Series	Code	Characteristics	Fan-in (UL)	Fan-out (UL)	Typical single gate propagation delay (D) (nsec)	Typical single gate power consumption (P) (mW)	Speed-power product (D×P) (pJ)
Standard	p74XXs	Basic series	1	10	10	10	100
Lower power Schottky	p74LSXXs	Schottky clamped to reduce power consumption	0·25	5·0	9·5	2	19
Advanced Schottky	p74ASXXs	Schottky design optimized for high speed	1·25	12·5	1·5	22·5	34
Advanced LP Schottky	p74ALSXXs	Improved LS series (lower power)	0·25	5·0	4	1	4
Fast	p74FXXs	Schottky design for high speed and medium power	0·5	12·5	3	12	36
High speed CMOS	p74HCTXXs	CMOS devices compatible with TTL 7400 series	0·05*	2·5	8	0·1*	1*
Advanced CMOS	p74ACTXXs	Improved HCT series	0·05*	15	5	0·1*	0·5*

*HCT and ACT, see text for fan-in; power consumption is proportional to frequency, value given is at 100 kHz.

Key to device codes:
p. A prefix unique to a manufacturer. (e.g. PC for Philips, SN for Texas Instruments, MC for Motorola).
s. A suffix which may vary from one manufacturer to another; it specifies the package (e.g. plastic or ceramic) and other mechanical features.
XX. A two, three, or four digit number which specifies the device function (e.g. 10 is triple 3-input NAND).
Examples: PC74HCT04P is a Philips HCT series, hextuple inverter in a plastic package.
SN74LS109AN is a Texas Instruments LS series dual JK bistable in a plastic package.

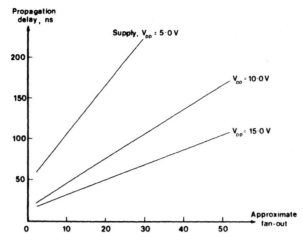

Fig. 7.3   CMOS element fan-out

obtained by examination of Fig. 7.3; this value should be slightly reduced (typically by about 10%) to allow for effects caused by the interconnection leads. Generally fan-outs above one hundred should be avoided.

4000 series and 74x00 series components must not be mixed. When it is necessary to connect one series to the other, special interface circuits must be built to convert the output signals from one into a form which meets the specification for input signals to the other. Also note that the reason for the low fan-in of the 74HCT00 series arises because these are CMOS devices, not TTL devices. However they have been constructed so that they may only be used with the normal TTL supply voltage of 5 V and their input and output circuits have the same voltage specifications as TTL devices.

Note that two circuit outputs must *never* be connected together; Boolean algebra does not define the result of such a connection and electronic logic elements are not designed to allow these connections. If such a connection is made the resulting output will be indeterminate and the devices may be permanently damaged.

7.3   Special interconnection techniques

It is convenient to assemble large systems from a number of separate independent modules or units, with the modules linked together by a large number of parallel connections. Usually modules may be plugged into, or removed from, this connection system; often they may be positioned at any point on the system and may use the parallel connections to send an output to, or to receive an input from, another module.

This assembly technique has the advantage that one-off quantities of special-purpose systems may be manufactured from standard units; each unit is therefore inexpensive and well tested. In addition any fault may be localized by replacing one module at a time until the fault is traced to a single unit.

The arrangement of parallel links between modules is termed a **bus** or a **highway** or a **dataway**. Obviously the rule that outputs must not be connected together will be broken by the bus arrangement; also the fan-out capability of an element may be exceeded since it is outside the control of the designer of the modules. There are techniques which allow a bus system to be implemented using standard components, but the most satisfactory method is to use special devices. One such device is the TTL gate with **'wired-OR'** capability; although these devices are still available as they have other uses a better solution is obtained using TTL or CMOS elements with **tri-state** outputs.

The name 'tri-state' is somewhat misleading; such devices do not have outputs which are able to take up three different levels. A device with a tri-state output behaves in the same way as a similar conventional logic element and its output may be 0 or 1. Additionally a special input to the device operates so that when one specified logic level is applied to this input the device behaves normally; when the opposite logic level is applied the tri-state element operates as though its output is disconnected. In this second disabled condition the device output appears to be an open circuit to anything connected to it. Therefore it may be freely connected to other outputs and to inputs without affecting them provided that it remains in this third (disabled) state. The input which controls the output condition is given many names; these include 'control', 'output control', 'enable' and 'disable'. There is no convention which defines the logic level required to enable the output; some devices require a 1 and others a 0. If the output is enabled when the control input is at 1 then the control is **active high**, if a 0 is required to enable the output the control is **active low**.

A system which incorporates a bus structure using tri-state elements usually operates in one of two ways; both schemes require a small number of control lines connected to all the units. In one system a particular module is a master unit and is the only one which is allowed to output signals to the control lines. Each module is allocated a different unique code corresponding to one combination of logic levels on the control lines and whenever the master unit outputs this code the unit may output to the common bus. In other words the tri-state outputs of a module may only be put into the active (enabled) condition when the control lines indicate that the module may output; the master unit is designed so that it only allows one unit to output at any time.

In the alternative scheme at least one control line is also driven by tri-state elements of every unit. In the most simple arrangement this line is connected to the logic 1 level through a resistor (pull up resistor). All units are of equal status and may output to the bus at any time, provided that the control line is at 1. To output to the bus a module examines the control line and if it is at 1 it enables its own tri-state output to the control line and forces the line to 0. Once a particular module has forced the control line to 0 it may output to the bus and hence to other units. When the output operation is complete the module disables all its tri-state outputs, including that to the control line, and the control line returns to 1. If a module attempts to output while another unit is using the bus it will find that the control line is 0 and must wait until the line becomes 1 before it can output.

Both techniques are usually implemented in a slightly more complex form to overcome problems which arise if two or more units request access to the bus simultaneously. In addition more advanced systems may include methods by which a unit sending signals to another unit may check that they have been correctly received.

Example 7.2

Design a circuit to connect the output of one of four remote data collection units to the single input of a data processing unit. The processing unit indicates the remote unit output required by a two bit code on two outputs.

Solution

Each remote unit signal is fed to its own tri-state buffer gate; these buffers are devices whose output, when enabled, is the same as the single input. The outputs of all the buffers are connected together and to the input of the processing unit. The two outputs from the processing unit are the inputs of the combinational circuit with Table 7.3 as its truth table. This circuit is a 2-line to 4-line decoder whose outputs are the active low tri-state controls of the four buffers. The complete system is shown in Fig. 7.4.

Table 7.3

Inputs		Outputs			
0	0	1	1	1	0
0	1	1	1	0	1
1	0	1	0	1	1
1	1	0	1	1	1

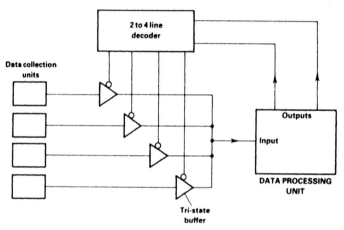

Fig. 7.4 An application of tri-state devices

7.4   External connections

Electronic circuits, no matter how large or sophisticated, are of no use by themselves. An electronic circuit will be used to perform some task and will frequently involve connections to non-electronic components. For instance a calculating device must have some method of receiving details of the calculations to be performed, and some way of indicating the results. In the most simple case the input will be from a keyboard (i.e. a set of switches) and the result will be indicated by a set of lamps. A more elaborate electronic logic device such as a controller for a machine tool may have to supply inputs to drive motors, relays, solenoids, etc., all of which must be operated by the electronic circuits.

In general, the input and output circuits (interface circuits) which connect an electronic logic circuit to external devices or to linear electronic circuits (i.e. non-logic ones) require a detailed knowledge of electronic circuit construction for their design.

Some integrated circuits are available for common interface tasks, but generally interface design requires experience and a knowledge of electronic circuit design techniques – although one or two simple interfacing problems may be solved using the circuits below.

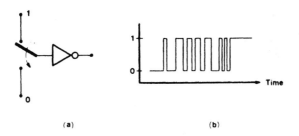

(a)   (b)

Fig. 7.5

It might appear to be reasonable to use the circuit in Fig. 7.5a to provide an input to a logic element (an inverter is illustrated) from a manually-operated switch. However this is not a satisfactory circuit; every time a mechanical switch is operated the spring material from which it is manufactured vibrates and the contacts open and close several times instead of making a single contact. The result of this multiple action is that the output of the logic element in Fig. 7.5a is similar to the waveform shown in Fig. 7.5b. This switch 'contact bounce' lasts for several milliseconds even with high quality switches, and in many applications the effects must be removed.

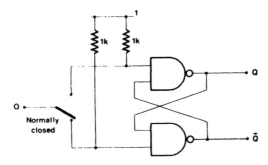

Fig. 7.6   Switch buffer circuit

Figure 7.6 shows one circuit which removes the problem, it is basically an $\overline{S}\,\overline{R}$ flip-flop with the $\overline{S}$ and $\overline{R}$ inputs connected to a spring-loaded changeover switch. The switch must be constructed such that one set of contacts open before the other set closes (break before make) and a single press and release operation produces a single 0 to 1 to 0 pulse at Q. There is a simultaneous 1 to 0 to 1 pulse at the $\overline{Q}$ output; in this application the circuit is termed a switch buffer circuit.

The design of simple output circuits is more difficult than the design of input ones. Energy is required to drive any device connected to an output and most non-electronic devices require more energy than standard logic integrated circuits can supply. Further, there are wide variations in the output capability of different series of logic devices; for example devices in some of the 7400 and 4000 series can only source (supply) about 0.4 mA when in the logic 1 state while others can source over 4 mA. An example of a simple output circuit is one which indicates the logic state at its input by turning a light-emitting diode (LED) on or off. Figure 7.7 is a suitable circuit; a high brightness type of LED must be used and it is on (illuminated) when input A is at logic 0, it is off when A is at logic 1. The transistor used is any general purpose small silicon planar type though 2N3904 is suggested as it is readily available and of low cost. Note the use of the electrical symbol for the inverter in Fig. 7.7; this is used as the circuit does not just show logic levels, it also includes electrical features.

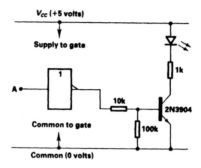

Fig. 7.7   7400 series output driving a LED

Another simple output circuit is one which provides the drive current to the coil of a relay. Slight modification of the circuit of Fig. 7.7 produces the version in Fig. 7.8 which has a relay coil as the collector load of the transistor. The relay must have a coil operating voltage of V_R which does not exceed about 25 volts and the relay coil resistance R_C must be such that the operating current does not exceed about 30 mA (i.e V_R divided by R_C is less than 30 mA). A general purpose low current diode is connected in reverse across the relay coil to protect the transistor. When the gate input B is at logic 0 current flows in the relay coil and it operates; no current flows when B is at logic 1. It is essential that the relay contacts driven by the coil are adequately rated for the task to which they are applied.

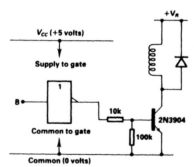

Fig. 7.8   Relay drive from a 7400 series device

7.5   General assembly points

Circuit connections should be short and neat; whenever possible leads should be well separated and, if possible, an earthed lead should be placed between adjacent signal leads. For simple systems which are not required to operate at maximum possible speed reasonable care will produce a reliable circuit. Large assemblies can only be constructed satisfactorily using printed circuit board (PCB) techniques; these must be used with high speed circuits and in such applications circuit layout requires knowledge of signal transmission techniques.

In addition to the power supply producing a steady output voltage at the required level, it is necessary to connect capacitors of about $0·1$ μF between the power supply leads close to each group of four or five integrated circuits. These decoupling capacitors should be a high-frequency type and are required because the internal construction of TTL and CMOS elements is such that large currents (relative to those normally required) flow when the output changes state. For TTL elements the currents are typically above 20 mA and of under 10 nanoseconds duration. These currents generate voltage spikes on the power supply lines; the capacitors remove these spikes and prevent one circuit element interfering with another through the power supply leads.

The final assembly point to consider is the manner of dealing with the unused inputs which arise in some circuit designs. These must always be connected to a suitable logic level and not left open circuit. This logic level may be one of the power supply lines but TTL inputs should be connected to a 1 level through a 1 kΩ resistor rather than directly connected.

7.6   Problems

1   Demonstrate that the circuit shown in Fig. 7.9 will generate a transient output when the inputs change from A = 1, B = 0, C = 1 to A = 1, B = 1, C = 1.

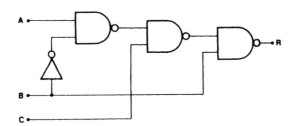

Fig. 7.9

Assume that all elements have an identical propagation delay and draw a timing diagram showing a change of 0 to 1 at the B input and the output transient on a common time-scale.

2   Devise a two-phase clock system and circuits which remove the transient output produced by the circuit in Fig. 7.9 examined in Problem 1.

3   Determine if the following connections will operate reliably (assume that all inputs and outputs have the normal fan-in and fan-out ratings for their own series).
a) A 74HCT series output driving two 74HCT inputs and three 74F inputs.
b) A 74F series output connected to four 74F inputs and four 74AS series inputs.
c) One 74LS output providing inputs to three 74AS and four 74F devices.
d) A 74ACT output driving two 74AS inputs, four 74ALS inputs and four 74F inputs.

4   Design a system utilizing tri-state output devices. The system has two control lines

which provide four different codes and a third line used for timing. These control lines define the action of a circuit which consists of four lines carrying logic signals connected to a single transmission line. The two code control lines identify one of the signal lines according to the code, and when the third line is in the logic 1 state, the line selected is connected to the transmission line so that signals on the selected line are transmitted.

8 Large logic networks

Earlier chapters describe reliable methods of designing combination and sequential logic networks. The designs produced may be implemented as assemblies of **small scale integrated circuits** (SSI devices); such devices have only a few basic logic gates or bistables in one integrated circuit package. When using SSI devices in a network the limits of real components, described in Chapter 7, must not be exceeded.

A major area of application of logic circuits is in systems which perform calculations; the fundamental operation required in these systems is to add together two binary integers each having the same number of digits (bits). The design of a multiple digit addition circuit involves many of the problems common to all large combinational networks and is examined in detail.

8.1 Addition of binary numbers

One example of the addition of two binary numbers is illustrated in Table 8.1a and the general case of the ith column of such an addition is shown in Table 8.1b.

Table 8.1a

Augend	1	1	1	1	0	0	1	1
Addend	0	1	1	0	0	1	1	0
Carry-in	1	1	0	0	1	1	0	–
Carry-out	1	1	1	0	0	1	1	0
Sum	1	0	1	0	1	1	0	0	1

Table 8.1b

		Column i
Augend		A_i
Addend		B_i
Carry-in		C_{i-1} ← Carry generated in column $i-1$
Sum	C_i	S_i

Carry to column $i+1$ (points to C_i)

Sum in column i (points to S_i)

8.1.1   Single digit addition

When two binary numbers A, the **augend**, and B, the **addend**, are added together Table 8.1b indicates that the result in the ith column will be a sum of S_i (0 or 1) in this column and a carry-out of C_i (also 0 or 1) to be added to the $i+1$th column. This result represents the arithmetical (not the Boolean) sum of the digits A_i, B_i, and C_{i-1} where A_i and B_i are the ith digits of A and B respectively and C_{i-1} is the carry-in which is generated as the carry-out from the previous $i-1$th column. If these binary digits are represented by Boolean variables and the addition is performed by a logic circuit which has inputs A_i, B_i, C_{i-1} and produces outputs of S_i and C_i, then Table 8.2 is the circuit truth table.

The carry-out of the ith stage has been labelled C_i; therefore the carry-in must be C_{i-1} as it is the carry-out from the previous, $i-1$th, stage. Some designers prefer to call the carry-in to the ith stage C_i and in this case its carry-out is C_{i+1}. The choice of which form to use is arbitrary as both are reasonable; however the two must never be used together and the form used must be clearly defined.

Table 8.2

Inputs			Outputs	
Carry-in	Digits		Sum	Carry
C_{i-1}	B_i	A_i	S_i	C_i
0	0	0	0	0
0	0	1	1	0
0	1	0	1	0
0	1	1	0	1
1	0	0	1	0
1	0	1	0	1
1	1	0	0	1
1	1	1	1	1

A Karnaugh map, or similar technique, could be used to derive Boolean expressions for S_i and C_i in terms of the inputs. However, the sum of products solution produced by such a technique is not the one which is most conveniently implemented in many cases; an alternative solution is obtained by the following two-stage treatment. Consider first the case in which the carry-in from the previous stage is not connected; i.e. only A_i and B_i are inputs to the addition circuit. Such a circuit is a **half-adder** and its outputs may be denoted by S_{i0} and C_{i0} where the subscript 0 indicates that there is no carry-in. Table 8.3

Table 8.3

Inputs		Outputs	
B_i	A_i	S_{i0}	C_{i0}
0	0	0	0
0	1	1	0
1	0	1	0
1	1	0	1

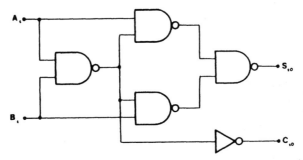

Fig. 8.1 Half-adder logic circuit

is the truth table for the half-adder and inspection gives the relationships

$$S_{i0} = A_i.\bar{B}_i + \bar{A}_i. + B_i = A_i \oplus B_i$$

$$C_{i0} = A_i.B_i$$

Figure 8.1 is the circuit diagram of a half-adder constructed from NAND gates and has S_{i0} and C_{i0} as outputs. The name half-adder is used because two half-adders may be connected to form a circuit which performs the complete addition with the carry-in from the previous stage connected. This complete circuit is a **full-adder** and may be described in terms of the half-adder outputs by the relationships

$$S_i = S_{i0} \oplus C_{i-1}$$

$$C_i = A_i.B_i + S_{i0}.C_{i-1} = C_{i0} + S_{i0}.C_{i-1} = \overline{\bar{C}_{i0}.(\overline{S_{i0}.C_{i-1}})}$$

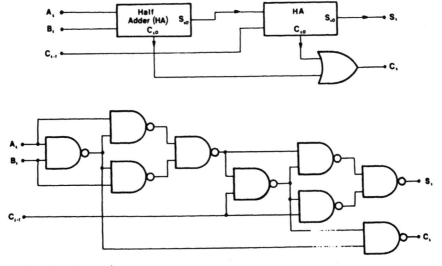

Fig. 8.2 Full-adder logic circuit

The circuit diagram for a full-adder constructed from half-adders and an expanded diagram with the half-adders implemented using 2-input NAND gates are both shown in Fig. 8.2. Many other logic circuits can be devised for both half and full-adders; the circuit shown is not the best when high speed operation is required. For highest speed the sum of products form (usually converted to all NAND or all NOR equivalents) should be adopted.

8.1.2   Multiple digit addition

A single full-adder only produces the result of the addition in a single column (i.e. the sum of one digit of each number and the carry-in) when the sum of two multiple digit numbers is required. However the full-adder is used as a component in most of the circuits which have been devised to add two multiple digit numbers. There are two general methods; in one a single full-adder is used and the digit pairs are added in turn, starting with the least significant pair. The other method is to have a separate full-adder for each pair of digits.

The technique which uses a single full-adder is known as **serial addition**, a five-digit version is shown schematically in Fig. 8.3. In this design each number is stored in a serial shift register, the carry-out is stored by a D-type master-slave flip-flop, and a control circuit ensures correct operation. The illustration shows no details of the control circuit and the augend register is used to store the result so that the original augend is 'lost'.

When the addition is performed by a circuit with a separate full-adder for each digit pair the circuit is a **parallel addition** circuit; a five-digit version is shown in Fig. 8.4. In such a circuit all the digits are input simultaneously, often by storing them in parallel in–parallel out (PIPO) registers. The carry-out from one full-adder is connected as the carry-in to the next; this circuit is often called a **ripple-carry adder**. The result appears at the sum outputs of all the adders.

Obviously a serial addition circuit is more economical than a parallel one in terms of the number of logic elements required but it will operate more slowly. Further, the control circuit of the serial adder is complicated. One fault of the parallel adder is the successive carry-out to carry-in connections; these connections generate long propagation delays. More elaborate circuits exist which overcome these long delays at the expense of added complexity.

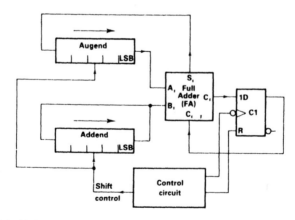

Fig. 8.3   Five-bit serial addition circuit

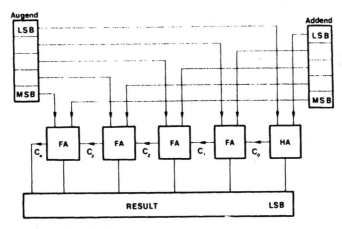

Fig. 8.4 Five-bit parallel addition circuit

Large electronic logic circuits can now be constructed so cheaply and operate with such high reliability that serial addition circuits are rarely used. Their main application is in pocket calculators, however the form used is not simple as the numbers are in BCD representation.

All other arithmetic circuits (i.e. subtractors, multipliers, etc.) may be based on addition circuits although most medium and large computers have special-purpose multiplication circuits.

8.2   The cell technique for combinational circuit design

The design of the parallel ripple-carry adder avoided the problems which arise when a direct approach is used to design a combinational logic circuit with a large number of inputs. The adder requires all the digits $A_0, A_1, A_2, \ldots, A_{n-1}$ of number A and all the digits $B_0, B_1, B_2, \ldots, B_{n-1}$ of number B as inputs; it has outputs $S_0, S_1, \ldots, S_{n-1}$ and C_{n-1}. That is a circuit to add two numbers each with n-bits must have $2n$ inputs and $n + 1$ outputs. For common applications n will be at least eight and often it will be much larger. Consequently design of the adder includes the problem of devising a circuit with a very large number of inputs and outputs. Although, as shown in Chapter 2, this can be reduced to separate single output circuits, the problem of designing circuits with many inputs remains. Obviously Karnaugh map techniques are not suitable and an alternative is required. It is relatively easy to program a computer to produce minimum Boolean expressions using the complete truth table as the program input. (Entry of the truth table can be a difficult task unless a program is devised to create it.) However the computer reduction does not assist significantly as, unless a great reduction is possible, any circuit constructed using the resulting expression will require logic gates with many inputs. For example, minimization, to sum of products form, of a circuit for multiple digit addition produces expressions for S_3 requiring 5-input AND gates and a 36-input OR gate. The number of OR gate inputs required to produce each S_i increases rapidly as i increases. Gates with many inputs can be manufactured but there is a finite limit to the number of inputs possible. The limit depends on the type of integrated circuit but is unlikely to exceed twenty, for many types it is significantly less.

The ripple-carry adder avoids the problems of deriving expressions for multiple input circuits and of obtaining multiple input gates for their construction. The adder is an example of a combinational logic circuit designed using the **cell technique**, it is not a formal method for which general rules can be devised. The method requires that the designer identifies a common block or sub-unit, a **cell**, which is easily designed and manufactured. The complete circuit is constructed using multiple copies of the cell as components. An extension is to allow a small number of different cell units to be used in the design, not just a single type.

The cell technique is very useful for the design of large combinational logic circuits, often it is the only suitable method, but it has disadvantages. Initially a choice must be made of the cell or cells; this requires experience combined with a process of trial and error. Secondly the resulting circuit will have significant propagation delays as a change in one output may cause many other gate outputs to change in a ripple through manner. In addition to the long delays, hazards, for example transients, may exist in the circuit.

Because the cell approach is commonly used for the design of large logic networks integrated circuit manufacturers provide many devices which are suitable for use as cells. These are functionally equivalent to a network requiring up to one hundred basic logic gates; devices of this complexity are **medium scale integrated circuits** (MSI devices). Examination of manufacturers' lists shows that a wide range of combinational and sequential MSI devices exists. Typical examples are the 74x83 4-bit adder and the 74x138 3-line to 8-line decoder.

8.3   Use of cells for large sequential circuits

The design of asynchronous divide-by-2^n and synchronous divide-by-N counters was described in Chapters 5 and 6. The general purpose divide-by-N counter design method of Chapter 6 becomes difficult to apply when N is large, also some circuits for the J and K control inputs require gates with many inputs. These problems may be avoided by using the cell technique to design sequential circuits although the method is not always suitable. For example it is usually difficult to devise cells when N does not have factors or the circuit has a large number of control inputs.

The most useful cells for sequential circuits are moderate size divide-by-N counters. Both synchronous and asynchronous counters are available as MSI devices and either type may be used as cells, generally synchronous types should be used if they are available. A simple method of constructing large counters is to use MSI counters as cells and connect them in an asynchronous cascade to form a counter of the required size. For example a divide-by-one-thousand counter may be built using three divide-by-ten synchronous counters as cells and connecting them as in Fig. 8.5a. Similarly a divide-by-sixty counter may be built using a divide-by-ten counter and a divide-by six counter as in Fig. 8.5b. Both circuits are only schematic (block diagrams, not circuit diagrams). They use the most significant bit output of each cell (counter module) as the input to the next cell and it is assumed that the cells are constructed using negative edge-triggered or master-slave bistables. (Minor design changes are required to use positive edge-triggered devices.) As for all asynchronous circuits the stages do not change simultaneously and transient, incorrect output states will occur.

When synchronous counters are built using cell techniques the cells themselves must be synchronous counters and must have a control input which can be used to prevent counting. That is, when the control input is in one logic state the counter remains in its present stage regardless of the clock input. A combinational circuit is used with each

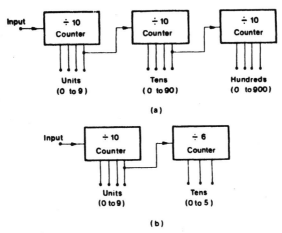

Fig. 8.5   Cascaded counter circuits

counter to provide a signal indicating when it is in the 'last state' (i.e. the next clock pulse will return it to the initial state) or when its control input is in the condition which prevents counting. Figure 8.6b shows the state diagram of a divide-by-six counter for use as such a module assuming that a control input, E, prevents counting when it is held at 0. Figure 8.6a is a circuit which behaves as the state diagram requires, it also includes the combinational circuit indicating 'last state'.

To build a large counter the combinational circuit output of one stage is used as the control input to the stage which immediately follows. The complete system is shown schematically in Fig. 8.7. This simple form has a long propagation delay path through the chain of combinational circuits which detect 'last state'; this causes problems when high speed operation is required. MSI devices are available which consist of a synchronous counter with control input and the combinational circuit to detect 'last state'. They have a slightly more complicated control mechanism to break the propagation delay chain and

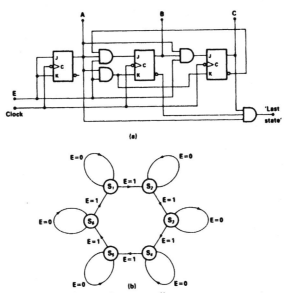

Fig. 8.6   Divide-by-six counter for use as a synchronous cell

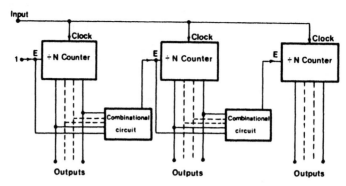

Fig. 8.7 Synchronous cascaded counter circuits

allow operation at high speed. The control requires two inputs and the application notes for such devices should be studied to determine the correct method of connecting them together to form large synchronous counters.

8.4   Other large sequential circuits

There are many methods of constructing counters in addition to those already described; many of the designs are based on shift registers. Generally the internal form of the shift registers is not important; the examples are drawn using edge-triggered D-type bistables. A simple shift register counter is the Johnson or 'twisted ring' counter; it is formed by connecting the final stage $\bar{Q}$ output as the first stage D input. The register must be set initially so that all the bistable Q outputs are zero and the pulses to be counted are applied to the register clock input. An n-stage Johnson counter will change through $2n$ different states; i.e. an n-stage Johnson counter is a divide-by-$2n$ counter. Figure 8.8 illustrates a five stage Johnson counter, Table 8.4 is its state table and illustrates its divide-by-ten action.

Table 8.4

| State | Shift register contents | | | | | Input to register on next shift |
| | LSB | | | | MSB | |
	A	B	C	D	E	$\bar{E}$
Initial, S_1	0	0	0	0	0	1
S_2	1	0	0	0	0	1
S_3	1	1	0	0	0	1
S_4	1	1	1	0	0	1
S_5	1	1	1	1	0	1
S_6	1	1	1	1	1	0
S_7	0	1	1	1	1	0
S_8	0	0	1	1	1	0
S_9	0	0	0	1	1	0
S_{10}	0	0	0	0	1	0
Initial, S_1	0	0	0	0	0	1

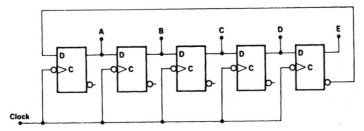

Fig. 8.8 Divide-by-ten Johnson counter

It is obvious that a Johnson counter requires more bistables than a counter designed using the method described in Chapter 6. The state sequence of Johnson counters is fixed and is not a binary counting sequence. There can only be $2n$ states, sequences with an odd number of states are not possible. The need to start with all zero (or any other combination of outputs in the count sequence) requires additional automatic power-on reset circuits in many applications. However this counter has the advantage that it is synchronous, requires no combinational logic to supply bistable controls, and is simple to design. Other forms of shift register based counter can be constructed; for example using combinational logic to produce the new register input as a function of the output of several stages can create sequences in excess of $2n$ states. The circuit given for Problem 6 in Chapter 4 is one such circuit. Some form of shift register based counter is often chosen when a counter is required as an internal element of an integrated circuit as this form is often the most compact.

Another method of designing large sequential circuits is to construct separate small circuits all driven synchronously from the same source of clock pulses. The state of one circuit is used to control the action of another; the state of this second counter is then used to control the first. The scheme may be extended and systems with several interconnected counters are possible.

Example 8.1

An industrial controller provides outputs controlling four process steps; the second step is active for fifty seconds, the other steps are each active for ten seconds. Design a suitable circuit using two counters driven by clock pulses at ten-second intervals to produce the control outputs.

Solution

There is a main counter with four states and a subsidiary counter with five states. When the main counter enters its second state, M_2, it enables the subsidiary counter which is already in its first state, S_1. The main counter remains in state M_2 until the subsidiary counter is in state S_5. The two counters control one another in a simple sequence.

If the main counter has bistable Q outputs A and B and state M_2 corresponds to $A = 1$, $B = 0$ then an AND gate with the output $G = A.\overline{B}$ produces the control for the subsidiary counter. (It is also the control output for the second process step.) Let the bistable outputs of the subsidiary counter be C, D and E with state S_5 having $C = D = 0$ and $E = 1$. A circuit generating $H = \overline{C}.\overline{D}.E$ will provide the control for the main counter. Figure 8.9 shows the state diagrams for the two counters and a simplified timing

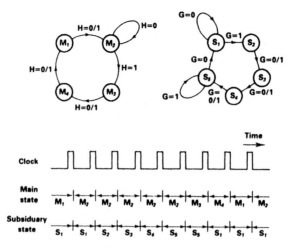

Fig. 8.9   State diagram and simplified timing diagram for Example 8.1

diagram. The detailed design of the counters is left to the reader as it is a simple exercise in the use of the techniques of Chapter 6. The four controller outputs are provided by separate gates indicating when the main counter is in each of its states (this section is the same as the scanner circuit of Section 5.4).

Example 8.1 has small numbers of states; in most situations an interconnected counter design would only be used for problems requiring a large number of states for which other techniques are difficult. The method may be extended to main sequences which have pauses at more than one position and the length of these pauses may differ. When this form of design is used care must be taken to ensure that the circuit cannot become stuck in some situation, this should not occur if the state diagrams are complete and show the effect of all combinations of all control inputs at every state. For complex multiple counter circuits a power-on reset circuit is essential to ensure a known starting point.

Whatever form of design is used a counter will often require combinational circuits between its outputs and the final circuit outputs; the design of these is straightforward and the scanner (Section 5.4) is again a simple example. Counters are frequently used in digital measuring instruments; examples include point of sale weighing machines, length measuring devices on machine tools, voltmeters, etc. Such instruments usually operate by converting the quantity being measured into a series of pulses whose number is proportional to the measurement. The pulses are input to a counter and the total count is displayed. In most applications the user does not wish to see the state of the counter while the count is in progress, only the final value is of interest. The usual method of producing a display which shows the total count is to use a display latch. This is a parallel in–parallel out (PIPO) register with the same number of stages as the counter. The outputs from all stages of the counter are connected as data inputs to the PIPO register, but while the count is in progress the clock input to the register is held so that no data is input. As soon as the count is complete a clock pulse is applied to the register and the total count is loaded into the PIPO register. The register outputs are connected to the display system which therefore always indicates the most recent total count.

8.5 LSI devices

For many applications integrated circuits containing large numbers of circuit elements are available. These are known as **large scale integrated circuits** (LSI devices); the largest devices are often called **very large scale integrated circuits** (VLSI devices). A wide range of LSI and VLSI devices is available for both specialist and general purpose applications and it is beyond the scope of this text to consider them all. However memory devices form a large group of LSI and VLSI devices, all have a similar simple basic organisation and are widely used.

A single bistable will store one binary digit (bit, logic state) and when the term memory is applied to a logic network it implies an organized arrangement of many storage elements. These storage elements are usually electronic bistable circuits although other two-state devices may be used; for example some memory components are based on magnetic effects.

In a logic memory the storage elements are grouped into blocks of equal size; these blocks are called **words** and typical word lengths (sizes) are 1, 2, 4, 8 and 16 elements (bits). Each word is identified by a number called the **address** of the word; the first word is at address zero (not one), the second is at one, the third at two and so on. The most useful memories allow reference to any single word by application of logic levels representing the binary number for the address of the word at address control inputs. Such memories are termed random access memories; an important feature is that the time to refer to a particular word (**access time**) does not depend on the address. An alternative form of memory is a sequential (or cyclic or serial) access one in which the user must start at address zero and refer to each word in turn; in such a memory the access time for a particular word depends upon the address.

Memories may be divided into two further classes; they are either **read only memories** (ROMs) or **read and write memories**. In the case of a ROM, information has been permanently written into the memory in some way and cannot be changed, i.e. the user can only obtain (read) the contents of any address. There are several types of ROM and each has a different method of initial loading. Some are loaded by a process in the original manufacture while alternative types, called 'programmable read only memories' (PROMs), are loaded by the user during an irreversible process.

Those memories which allow the user to both read the contents of any location and enter new values (write) into any location are read and write memories but are known as RAMs (**random access memories**). This is confusing because most ROMs are also random access; normal use of ROM implies random access read only memory while RAM implies random access read and write memory.

In general, a RAM is organized as shown schematically in Fig. 8.10. A ROM is arranged in a similar manner but there is no write (data input) facility. To enter information into a particular word of a RAM, logic values which represent the address of the word in binary form are applied to the address inputs. Simultaneously, the new data is supplied at the data inputs and a pulse is input to the write control. The contents of a word of a RAM or ROM are obtained by input of the address as in a write operation and application of a pulse to the read control; the contents of the word appear at the data outputs. Many designs of RAM use a tri-state bus system for the data and have a single set of data connections instead of separate input and output connections.

Memories are available with several million storage elements in a single integrated circuit; one simple application of memories is to use a ROM in place of a complex

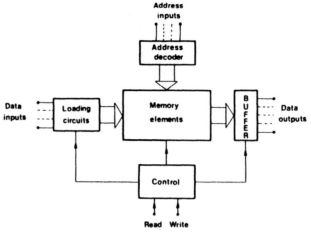

Fig. 8.10   Organization of a random access memory

multiple-output combinational logic circuit. When a ROM is used in this way the address inputs are the combinational circuit inputs, the read control is arranged so that the ROM always outputs the contents of the currently selected address, and the data outputs are the circuit outputs. Circuit design simply consists of ensuring that the ROM contains the information which causes the circuit to behave as required by the truth table.

Two common applications of ROMs as circuits are for code conversion and arithmetic operations, especially multiplication. Figure 8.11 illustrates the use of a ROM to convert a three bit pure binary code into one particular Gray code. In this example the conversion could simply be performed by a combinational circuit. However the ROM design is easily extended to more realistic cases of codes with many bits whereas direct combinational design becomes difficult when the number of bits is large.

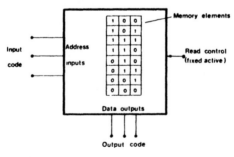

Fig. 8.11   ROM based code converter

8.6   Comments

Several techniques have been described for the design of large logic circuits. Most techniques require that the large circuit is broken into a number of smaller units; the units are designed using standard techniques and are interconnected to form the complete circuit. There is no formal design method for the division of a large circuit into small units; the process generally relies on the experience of the designer although a study of existing designs can assist by suggesting approaches to particular problems. In all cases care must be taken to ensure that hazards are not created in the design process or, if they are unavoidable (for example transients in the ripple-carry adder), they are not allowed to affect any circuits which use the outputs.

8.7 Problems

1 Devise a circuit to subtract two 8-bit numbers using the cell technique.

2 The parity of a binary number is defined as even if the number of ones in the number is even and is odd if the number of ones is odd. Design a circuit to indicate the parity of an 8-bit number, the circuit is to have a single output which is 1 when the parity is odd and is 0 when the parity is even.

3 Design a divide-by-eight counter with control input which inhibits counting and an output which indicates 'last state'. Use several counters of this design as modules to design a totally synchronous divide-by-4096 counter.

4 Using a clock with a one second period devise a circuit with two, or more, counters to act as the control system for a set of traffic signals at a simple crossroads junction.

   The main counter should have eight states corresponding to each of the signal conditions (UK sequence) with intermediate states (yellow signal lamp on) existing for one second. The signals for both roads are simultaneously red for five seconds and the signal for each road is green for fifteen seconds.

9 Application specific integrated circuits (ASICs)

Large logic networks may be constructed as assemblies of readily available integrated circuits; the exceptionally high reliability of these devices ensures that networks containing large numbers of SSI, MSI and LSI devices will function correctly. Often the cost of assembling a network forms a significant part of the total manufacturing cost (including purchase of the integrated circuits), so that the use of a small number of LSI devices instead of a large number of SSI and MSI devices may reduce total costs. Also a reduction in the number of components usually increases the, already high, system reliability. Consequently designers try to minimize the number of integrated circuits required in large networks by using large scale devices. However, as most new network designs implement functions for which LSI devices are not available, the use of a small number of devices may require the manufacture of a unique LSI device for a user's special requirement. Such an integrated circuit is an **Application Specific Integrated Circuit (ASIC)**. Although integrated circuit manufacturing costs are low when many identical devices are made, the cost of preparing a new design is extremely high. Therefore, unless production quantities will be high, designers of large logic networks sometimes have to choose between low assembly costs using expensive ASICs and high assembly costs using available SSI and MSI devices.

To overcome the high cost of ASIC production manufacturers of integrated circuits have designed several forms of device which allow ASICs to be used in many applications; not just those requiring large quantity production.

9.1 Approaches to ASIC manufacture

If a product containing a logic network is to be made in large quantities a special purpose LSI device can be designed and manufactured using standard integrated circuit production techniques. In general the large overhead costs only allow this approach to be used when tens of thousands of devices are required. ASICs of this type are called **full custom integrated circuits** and the semiconductor manufacturer usually undertakes their detailed design. When the required quantities are high, but not sufficient to justify full custom design, an alternative approach is one in which only a small part of the manufacturing process is unique for an application. Devices of this form are known as **semi-custom** devices and many methods exist for their production.

A typical example of a semi-custom device is the uncommitted logic array (UCLA). Integrated circuits consist of a large number of circuit elements (transistors, resistors, etc.) formed on the surface of a piece of semiconductor material in a sequence of about ten production operations. One of these operations connects the components to form the required circuit by depositing conducting tracks in a single process with the pattern of tracks set by a mask (similar to a photographic negative). To make an UCLA the manufacturer designs an integrated circuit consisting of an array of a large number of basic logic gates, bistables, etc. This is manufactured with the same array used for all circuits, only the interconnection pattern joining the gates to form the required network is unique for an application. That is, only the mask used to form the gate interconnections and the final step in the production process differs for each ASIC design. Consequently costs are lower than for a full custom design and relatively small quantities can be produced economically. There are disadvantages which arise mainly because the circuit has to be made to 'fit' the number, range and position of gates on the UCLA. However, when moderate production quantities are required semi-custom methods are often the most suitable ones.

Full and semi-custom ASICs are rarely economical when only small quantities of a particular design are required. To allow ASICs to be more widely used, integrated circuit manufacturers have developed methods which allow the user to form the required circuit after all the manufacturing processes, including encapsulation of the device, are complete. These devices have logic structures similar to those of many semi-custom devices but are on a smaller scale (fewer logic gates). A wide range of possible connections is built into into the device in manufacture, then, depending on the internal mechanism, the user either makes or disconnects some of the connections in a process known as programming. In general the extra circuits required to perform the programming operation mean that, in terms of logic complexity, the devices are smaller than full custom or semi-custom types. However, production of the required design is under the control of the user rather than the manufacturer, and the circuit can be produced and used immediately after the design has been completed.

Only user-programmed ASICs are described in further detail as, in most cases, these are the only forms that can be used for exercises that extend from devising the specification, to testing and using the final large network.

9.2 User programmed ASICs

ASICs that are programmed by the user are available in several forms. Some types may only be programmed once as the process is permanent. If an error is made in the design, or if the specification is modified, any programmed devices must be discarded and new ones used. These devices usually have the circuit connections completed by fuse links and are manufactured with all possible links completed. In the programming process the user destroys selected links leaving complete the links necessary to form the required circuit.

There are several types of device which have a reversible programming process. In these the connecting links are made by transistors which act as switches and each switch has its own control transistor. The control transistors are MOS types with a special construction for which there are several proprietary names, FAMOS being the most common one. This transistor has the property that it can set the switch it controls to be open or closed and it holds the switch in the position chosen even when power is removed. All the switches may be returned to the initial, unprogrammed condition by a process called erasure. Most devices are erased by exposure to ultra-violet (UV) light for several minutes: these have the obvious feature of a quartz window in the package. There are also devices which are erased by electrical signals, the extra complexity of these makes them more expensive and fewer types are available.

All user-programmed logic devices are **programmable logic devices (PLDs)** but manufacturers also use proprietary names (PAL, PLA and others). The devices that can be erased are erasable programmable logic devices (**EPLDs**), and may be further classified by the erasure process as uvEPLDs or EEPLDs. It might appear that the fuse link type of device is not required as the reprogrammable types have the advantage that they can be re-used if an error is made or the specification is changed. However each different type of device has properties making it suitable for certain applications. In general the fuse link devices are built using bipolar transistors and can be operated at much higher speeds than the erasable types which are built using MOS transistors. However, much larger scale circuits can be constructed using the MOS devices and the circuits produced use less power. Usually the fuse link devices are only used when their high speed is required, in other cases one of the erasable types is chosen.

9.3   Combinational circuits

A method of using a ROM as a combinational logic circuit was described in Chapter 8. ROMs are one particular form of ASIC, they are readily available in semi-custom forms, in once only programmable, and in reprogrammable forms. In many cases a ROM is not an economical method of producing a combinational circuit as the whole truth table must be stored in the ROM, thus for a circuit with n inputs and m outputs $2^n \times m$ memory cells are required. When it is possible to produce simplified Boolean expressions for a circuit the number of logic elements required to construct the circuit is much less than $2^n \times m$. A further disadvantage of very large ROMs is that they tend to have long propagation delays. However ROMs have the advantage that circuit design is very simple as the truth table for the required circuit is the programming pattern for the ROM.

For most applications which require a user-programmed ASIC as a combinational circuit the most suitable device is a PLD. These are available with many internal structures, most are based on the sum of products forms of Boolean expression for combinational circuits described in Chapter 3.

In general a user-programmed device contains a section known as the **matrix** or **programming matrix**; this consists of two sets of conductors which are usually shown on diagrams as sets of parallel lines intersecting at right angles. The section containing the large number of conductor crossing points is the matrix. One of the conductor sets forms the inputs to the matrix, the other set are the outputs from it. Connections are made or broken at each matrix crossing point during the programming process. Each external input to a PLD based on sum of products form produces two matrix input lines; one line is supplied directly (or through a non-inverting buffer circuit) and the other line is supplied through an inverter. Therefore in the matrix, as outlined in Fig. 9.1, the true and complement values of every input are available. Usually the triangle symbol with two outputs, one from a circle showing inversion, is used to show the generation of the buffered true and inverted matrix inputs. This symbol should not be confused with that for a tri-state buffer, it differs slightly. The symbol may also be drawn in several slightly different forms.

A PLD contains a large number of AND gates and the outputs from the matrix are used as the inputs to these gates. In the programming process every true and every inverted input in the matrix may be connected as an input to any number of AND gates. Almost all types of PLD allow every matrix input to be used as an input to every AND gate; that is, all the AND gates in an n-input device have $2n$ inputs. AND gate inputs which are left unconnected after programming are arranged to be at the logic 1 level.

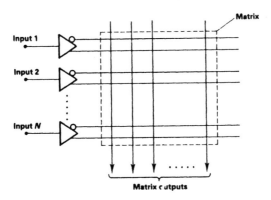

Fig. 9.1 Schematic representation of the matrix of a PLD

Therefore programming both the true and inverse of the same input to an AND gate will force its output to be permanently 0. Programming no inputs to a gate will force its output to be permanently 1. Because there are many AND gates in a PLD, each with many inputs, diagrams of the matrix do not show the separate inputs to the AND gates. Instead a single output line from the matrix is drawn as the only input to each AND gate. When a device is drawn in the programmed state a cross is put on the AND gate input line where it crosses each matrix input line which has been used as an input to the AND gate. For example Fig. 9.2 shows a device with inputs A, B and C with four AND gates producing the values $A.B.\bar{C}$, $\bar{A}.B$, $\bar{B}.\bar{C}$ and $\bar{B}$ at their outputs. If a dot is shown where two lines meet in the matrix it has the normal meaning and indicates a permanent connection.

Figure 9.2 is only a partial diagram, it does not include the OR gates required to produce the sum of products result. Some types of programmable logic device have the AND gate outputs as inputs to a second programmable matrix but most types do not, only those *without* the second matrix are described here. The manufacturers of PLDs offer ranges of devices with differing numbers of inputs and outputs; selection of the most suitable device for an application allows the proportion of unused gates to be kept relatively low. Each final output from the PLD has its own OR gate with several AND gates connected to provide the inputs to the OR gate; some devices have identical numbers of AND gates for every OR gate while others have varying numbers. Figure 9.3 illustrates a typical device but the numbers of inputs, outputs and AND gates shown are much lower than those for any actual PLD. The device in Fig. 9.3 is shown programmed

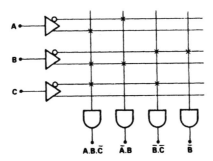

Fig. 9.2 AND term generation by a PLD

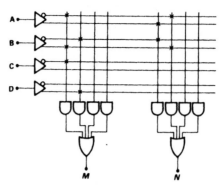

Fig. 9.3   PLD programmed to produce $M = \bar{A}.B.\bar{C} + \bar{B}.D$ and $N = A.\bar{B} + \bar{A}.B$

to produce $\bar{A}.B.\bar{C} + \bar{B}.D$ on output M and $A.\bar{B} + \bar{A}.B$ on output N.

To use a PLD the appropriate connections in the matrix must be made or broken (depending on the type an unprogrammed device will either have all links open or all links closed). The user does not need to consider which connections to make in detail as, to encourage the use of the devices, manufacturers provide a large amount of computer aided design (CAD) material. The material varies in complexity and quality from one manufacturer to another but all the systems include methods which enable the designer to describe the required circuit. Methods of description that are used inlude truth tables, Boolean expressions, and logic circuit diagrams prepared on the computer screen using a graphics package. Regardless of the input method used, the CAD package usually converts the description of the circuit into a Boolean sum of products expression and minimizes this into the most suitable form for the type of device being used. A fitting program then determines how to program the device to perform as required, that is it determines which links must be made and which must be broken in the matrix. The output from the fitting program is a file of instructions for a unit called a programmer; the user inserts a blank (unprogrammed or erased) PLD into the programmer, gives a simple start command and the programmer automatically creates the required link pattern using the instruction file.

The combinational PLD in Fig. 9.3 is the most simple structure used; each output can be used to implement any combinational circuit which is a function of some, or all, of the inputs. The only restriction is the number of AND gates connected to each OR gate; if the minimum sum of products expression for the circuit has more product terms than there are AND gates then the circuit will not 'fit' the device. If this occurs a larger device must be selected or some form of cell technique (with the usual faults) may be used with each AND–OR group used as a cell.

Many PLDs are more complex than the form in Fig. 9.3; a common additional feature is a tri-state buffer between the OR gate and the output pin. This buffer often performs logic inversion in addition to providing the tri-state function. The inversion does not cause problems as any combinational circuit may be described either by a sum of products expression or by an inverted sum of products expression. For example if an output of $\bar{A}.\bar{B}.\bar{C} + \bar{A}.B.C + A.\bar{B}.C + A.B.\bar{C}$ is required then a device which has an inverting output buffer must be programmed to produce

$$\overline{\bar{A}.\bar{B}.C + \bar{A}.B.\bar{C} + A.\bar{B}.\bar{C} + A.B.C}$$

which is the dual re-arranged and then inverted by the buffer. There are occasions when a circuit will fit a device which has an inverting buffer but will not fit a device with a non-inverting buffer. There are an equal number of cases when the opposite is true. Depending on the PLD used the enable control of the output buffer may be directly from an input pin or it may be supplied by an AND gate which has inputs from the matrix. In some devices all output buffers operate with their enable controls connected in parallel, other types have a separate control for each buffer. When the tri-state output is controlled by an AND gate the output pin may also be connected into the matrix and the true and inverse values of the output are available as AND gate inputs.

Figure 9.4 is diagram showing the structure of a small part of an unprogrammed PAL16L8 device which is one of the smallest PLDs available. The device has ten inputs and eight OR gates, each OR gate has its own tri-state buffer and output pin but only six of the outputs are connected back into the matrix. The connection of the output into the matrix converts the PLD output pin into a combined input and output pin. If the tri-state buffer is permanently disabled by programming the pin acts as an extra input (at the expense of leaving the associated AND–OR section unused). When the tri-state buffer is permanently enabled the output is always available in the matrix and larger circuits can be constructed using cell techniques without having to make the cell-to-cell connections externally.

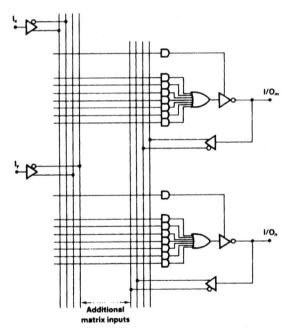

Fig. 9.4 Part of an unprogrammed PAL16L8 device

9.4 Sequential circuits

Sequential circuits are usually more complicated than combinational ones, consequently a more diverse range of PLD forms is available to implement them and the devices are more complex than those intended only for combinational circuits. It is possible to use the connection of buffer outputs into the matrix, as in Fig. 9.4, to create feedback and hence form a sequential circuit. This is not recommended as in most applications this does not use the elements of the PLD economically. When a sequential circuit is constructed with a PLD, a device intended for sequential use should be chosen.

The most simple forms of PLDs for sequential circuits have bistables in the section of the device associated with each output. The range of possible structures is very large and no attempt is made to review them here. The devices are complex so it is not customary to show the whole circuit of a device in detail. Instead, each OR gate, its input AND gates, the section of the matrix associated with these AND gates together with the output, including any bistable, are regarded as a complete unit; this unit is usually called a **macrocell**. Each different PLD has of a number of macrocells (usually all identical) and can be described by a diagram using blocks to represent each macrocell with a separate diagram showing the macrocell structure. Figure 9.5 outlines a typical form for a macrocell, and is a relatively simple macrocell as the bistable section consists only of a positive edge-triggered D-type flip-flop. Figure 9.6 shows a device containing four macrocells.

There are many more features to be selected when determining how to program a PLD suitable for implementing sequential circuits than when using one intended only for combinational circuits. Many sections of the macrocell contain programmable elements, that is in addition to the matrix supplying the AND gates there are many other programmed links which must be set to the correct state to form the required circuit. The following list describes some of the features which may be selected when programming each macrocell; not all PLDs will have all the features and most will have additional features.

a)   The clock of the bistable may be obtained from the common PLD system clock line. All clocks connected to this line are in parallel and bistables connected this way form a synchronous sequential circuit. Alternatively the bistable clock may be derived from an AND gate which has its inputs supplied by the matrix; in this case the bistable forms part of an asynchronous circuit.

b)   Most large circuits require some sections which are only combinational. Programming of the macrocell element barked SW3 in Fig. 9.5 allows the output from the OR gate to be arranged to by-pass the bistable. This produces a simple combinational function.

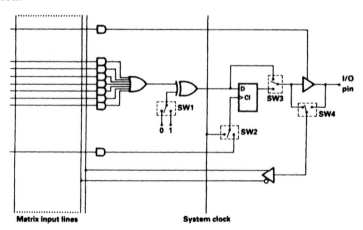

Fig. 9.5   A typical macrocell

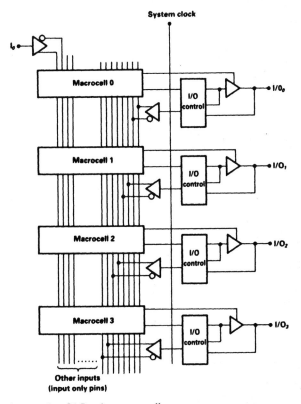

Fig. 9.6   Organization of a complete PLD using macrocells

c)   Figure 9.5 shows the OR gate output as one input to an exclusive-OR gate, the other exclusive-OR input may be programmed to be 0 or 1. When this input is programmed as 0 the gate has no effect (signal out equals signal in) but when it is programmed as 1 the signal from the OR gate is inverted. This optional inversion can assist in fitting a function into the device. For example, the expressions $A.B+A.\bar{C}+B.\bar{C}+\bar{A}.\bar{B}.C$ and $\overline{\bar{A}.\bar{B}.\bar{C}}+\overline{\bar{A}.B.C}+A.\bar{B}.C$ are identical, but the inverted expression requires one AND gate less than the one which is not inverted. For most PLDs the number of AND gates available as inputs to each OR gate is limited but the number of inputs to each AND gate is not restricted. Therefore the process of Boolean minimization required to fit a design into a PLD is not that of producing a minimum Boolean expression, it is one of minimizing the number of AND gates required. The optional inversion assists in this minimization.

d)   The feedback to the matrix may be obtained by programming the element SW4 in Fig. 9.5 so that it is obtained before or after the tri-state buffer. This feature can assist in fitting certain types of circuit into the device.

e)   Although not shown in Fig. 9.5 many PLDs allow the bistable to be programmed to function as any clocked type (i.e. D, JK, clocked SR, or T). When the bistable requires two control inputs the OR gate is modified to form two OR gates with the AND gates shared between them. The ability to select the type bistable makes the decisions regarding how to program the device more complicated but allows much more efficient use to be made of the PLD elements (larger circuits can be fitted into a PLD of a particular size).

It should be apparent from the range of features described that PLDs for sequential circuits are complex; even small low cost devices often have so many macrocells that the macrocells themselves are organized into groups. The only reasonable method of using these devices to implement logic circuits is with the aid of the computer support packages supplied by the device manufacturers. Packages for sequential circuit design are similar to those for combinational design but also include methods of describing sequential circuits by input of the circuit's state diagram or state table. Once the circuit has been specified the support program determines the most suitable method to implement the circuit using the device selected (some packages will determine the smallest device which can be used to implement the circuit) and determines the programming pattern.

9.5   Concluding remarks

This book has described some basic methods which may be used to design logic circuits. A few of the techniques that are used when circuits become too large for direct application of the basic methods have been introduced and more advanced treatment of most topics may be found in the texts listed in the Bibliography.

When a logic circuit is used as part of a product it will normally be manufactured using ASICs (unless the circuit is very simple). Although the computer programs which assist in the design of ASICs reduce the design effort required, it is essential to understand the basic methods of logic circuit design to use these programs effectively. Expertise is necessary to select the most suitable device and to understand possible hazards. Also the CAD programs generate warning and failure messages because the design (as specified) has faults or is not compatible with the PLD selected. These messages require that the user determines and understands their cause then modifies the specification or choice of device to avoid the particular problem.

Appendix A

Number Systems and Base Conversions

Number systems and number representation using a fixed base were described in Chapter 1. The use of a base of ten gives the familiar decimal number system which requires only ten different symbols to represent all possible numbers. Usually the symbols chosen are the printed ones 0, 1, 2, 3, 4, 5, 6, 7, 8 and 9; it is assumed that readers are familiar with decimal numbers and are able to manipulate them as required to perform calculations. When the signals in logic networks are used to represent numeric values the numbers are usually in binary form, that is, the value of the base is two. Binary numbers require only two digit symbols to represent a single unit (one) and nothing (zero, or null). Generally the symbols 1 and 0 are chosen for the digits of printed binary numbers as they are already familiar from their use in the decimal (denary) system.

An example of a binary number is $101 11011 \cdot 11001_2$ where the subscript 2 is used to indicate that this is a binary number; without the subscript the number could be one with any base. Using decimal notation and values for the powers of two this number is equivalent to

$$1 \times 2^7 + 0 \times 2^6 + 1 \times 2^5 + 1 \times 2^4 + 1 \times 2^3 + 0 \times 2^2 + 1 \times 2^1$$
$$+ 1 \times 2^0 + 1 \times 2^{-1} + 1 \times 2^{-2} + 0 \times 2^{-3} + 0 \times 2^{-4} + 1 \times 2^{-5}$$

which is more conveniently written as

$$128 + 0 + 32 + 16 + 8 + 0 + 2 + 1 + 0 \cdot 5 + 0 \cdot 25 + 0 + 0 + 0 \cdot 03125$$

giving $187 \cdot 78125_{10}$ as the total. Here the subscript 10 indicates that the printed number is a decimal number (base ten).

A.1 Conversions between the binary and decimal systems

Provided that numbers to be converted do not have either a very large or a very small magnitude then a simple method of converting a binary number into its decimal equivalent is the technique already illustrated in which each digit is multiplied by the weight, 2^n, expressed as a decimal number. The value of the weight is set by the digit position relative to that of the point. This technique is simple because all the calculations use well known rules of the familiar decimal system.

The reverse process of converting decimal numbers to binary numbers is not so obvious; the conversion is most easily performed using the following technique which is a particular case of a general method for the conversion of a number in one base into its equivalent in another base. Before conversion the decimal number is separated into integral and fractional parts which are converted separately.

Conversion of the integral part is performed by dividing it by two to produce an integral result and a remainder. The remainder is noted and the integral result is divided by two to give a new integer result and another remainder which is also noted. The process of dividing by two and recording the remainder is repeated until the integer result is zero, i.e. the last stage is always $1 \div 2 = 0$ remainder 1. The binary equivalent of the original decimal integer is the remainders in reverse order; that is, the last remainder produced is the first (the most significant) binary digit. Table A.1 illustrates the procedure for the conversion of the decimal integer 363_{10} into its binary equivalent. The remainders written in the reverse order of generation give 101101011_2 as the binary equivalent of 363 decimal; this may be checked using the previous expansion technique to convert 101101011_2 to decimal.

$$1 \times 2^8 + 0 \times 2^7 + 1 \times 2^6 + 1 \times 2^5 + 0 \times 2^4 + 1 \times 2^3 + 0 \times 2^2 + 1 \times 2^1 + 1 \times 2^0$$

is $1 \times 256 + 0 + 1 \times 64 + 1 \times 32 + 0 + 1 \times 8 + 0 + 1 \times 2 + 1 \times 1$ which equals 363_{10}.

Table A.1

$363 \div 2 = 181$	remainder, R = 1
$181 \div 2 = 90$	remainder, R = 1
$90 \div 2 = 45$	remainder, R = 0
$45 \div 2 = 22$	remainder, R = 1
$22 \div 2 = 11$	remainder, R = 0
$11 \div 2 = 5$	remainder, R = 1
$5 \div 2 = 2$	remainder, R = 1
$2 \div 2 = 1$	remainder, R = 0
$1 \div 2 = 0$	remainder, R = 1

A common error made when performing decimal to binary conversions is to stop when the integer result is 1 instead of continuing until it is zero.

The fractional part of a decimal number is converted to its binary equivalent by repeated multiplication by two. The integer digit (always 1 or 0) produced to the left of the point by each multiplication is recorded but only the fractional part of the result is multiplied in the following stage. The integer digits recorded form the required binary fraction when they are written in the order in which they were produced (i.e. the first integer produced is the digit immediately to the right of the point). The process is

Table A.2

$0.413 \times 2 = 0.826$	integer, I = 0
$0.826 \times 2 = 1.652$	integer, I = 1
$0.652 \times 2 = 1.304$	integer, I = 1
$0.304 \times 2 = 0.608$	integer, I = 0
$0.608 \times 2 = 1.216$	integer, I = 1
$0.216 \times 2 = 0.432$	integer, I = 0
$0.432 \times 2 = 0.864$	integer, I = 0
$0.864 \times 2 = 1.728$	integer, I = 1
$0.728 \times 2 = 1.456$	integer, I = 1
$0.456 \times 2 = 0.912$	integer, I = 0
etc.	

illustrated for the conversion of the decimal fraction 0.413_{10} in Table A.2. In general the process of conversion continues indefinitely; a decimal fraction rarely corresponds to a finite length binary fraction. When the conversion is not exact it must be stopped at some stage and the truncated result must be rounded to the correct value. In the example, the value of 413_{10} is equivalent to 0.011010011_2 to nine binary places, but to eight places the correct (nearest) value is 0.01101010_2.

Combining the results of the two examples gives 101101011.011010011_2 as the binary equivalent of the decimal number 363.413_{10} to nine binary places.

A.2 Other bases

Although any base may be used to represent numbers, ten is the base commonly used to perform calculations manually. In computers and related systems base two is used. However, many digits are required for moderate magnitude numbers in binary form and consequently octal (base eight) or hexadecimal (base sixteen) representations are often used to display numbers held in binary form by computers. Base twelve (the duodecimal system) and base sixty are used in some systems (12×2 hours in a day, $5 \times 12 = 60$ seconds in a minute, 6×60 degrees in a circle). These uses of base twelve (the duodecimal system) are the remnants of some very early number systems.

As each base requires a number of digit symbols equal to the value of the base some additional symbols are required if bases above ten are used. The normal choice is to use the letters of the alphabet as the extra symbols with A for ten, B for eleven, and so on. Table A.3 shows the printed forms of the first twenty numbers in several bases.

Table A.3

Value	Binary	Octal	Decimal	Duodecimal	Hexadecimal
zero	0	0	0	0	0
one	1	1	1	1	1
two	10	2	2	2	2
three	11	3	3	3	3
four	100	4	4	4	4
five	101	5	5	5	5
six	110	6	6	6	6
seven	111	7	7	7	7
eight	1000	10	8	8	8
nine	1001	11	9	9	9
ten	1010	12	10	A	A
eleven	1011	13	11	B	B
twelve	1100	14	12	10	C
thirteen	1101	15	13	11	D
fourteen	1110	16	14	12	E
fifteen	1111	17	15	13	F
sixteen	10000	20	16	14	10
seventeen	10001	21	17	15	11
eighteen	10010	22	18	16	12
nineteen	10011	23	19	17	13

A.3 Conversions between bases

The methods used to convert decimal integers and fractions into their binary equivalents may be adapted to form general methods for the conversion of a number in one base into its equivalent in another base. Conversion of the integer part requires a repeated division sequence. Working with all values in the original base, repeatedly divide the integer value by the required new base recording the remainders exactly as for the decimal to binary conversion (except that remainders will have any value from 0 to the new base minus one). Change the remainders into the digit symbols of the new base and write them in the reverse order of production (the last remainder produced as the most significant digit). This is the equivalent integer in the new base. The repeated multiplication method used to convert decimal fractions into binary ones may be adapted in a similar manner.

Although the general methods can always be used, conversion is particularly simple when one base is a simple power of the other. Thus conversions between binary and octal (2^3), or between binary and hexadecimal (2^4) are very easy. It is often convenient to convert to or from bases eight and sixteen via a binary intermediate value as binary manipulations are simple and the other conversion can be performed by inspection.

Converting an octal number into its binary equivalent only requires that each octal digit is replaced by a three-digit binary equivalent. That is to replace octal digit 2 the binary pattern 010 must be used; note that the leading 0 must not be ignored.

For the octal number $425 \cdot 713_8$ the conversion is

$$
\begin{array}{ccccccc}
4 & 2 & 5 & \cdot & 7 & 1 & 3 \\
100 & 010 & 101 & \cdot & 111 & 001 & 011
\end{array}
$$

Thus $100010101 \cdot 111001011_2$ is the binary equivalent of $425 \cdot 713_8$.

The reverse conversion is equally simple; starting from the point separate the binary number into groups of three digits. It is essential to start at the point and divide the integer and fractional parts separately. If the group at either end of the number does not have three digits, add extra zeros to complete the group without changing the binary value.

The conversion of $10111011 \cdot 11001_2$ into octal is illustrated below.

$$
\begin{array}{ccccccc}
010 & 111 & 011 & \cdot & 110 & 010 \\
2 & 7 & 3 & \cdot & 6 & 2
\end{array}
$$

Thus $10111011 \cdot 11001_2$ is equivalent to $273 \cdot 62_8$.

Conversions between base two and base sixteen use the same method as those between bases two and eight except that groups of four binary digits must be used. For example $3EC \cdot 2F8_{16}$ is converted as follows.

$$
\begin{array}{ccccccc}
3 & E & C & \cdot & 2 & F & 8 \\
0011 & 1110 & 1100 & \cdot & 0010 & 1111 & 1000
\end{array}
$$

Thus $1111101100 \cdot 001011111_2$ is the binary equivalent.

Binary to hexadecimal conversion of $10110101101 \cdot 0110101101_2$ is

$$
\begin{array}{ccccccc}
0101 & 1010 & 1101 & \cdot & 0110 & 1011 & 0100 \\
5 & A & D & \cdot & 6 & B & 4
\end{array}
$$

giving a result of $5AD \cdot 6B4_{16}$. Note that when one base is a simple integral power of the other, conversion of the fractional part is always exact.

Appendix B

Maxterm Representation of Circuits

In Chapter 3 a minterm representation was developed for combinational logic circuits and, as in all techniques involving Boolean variables, there is an equivalent dual method. This dual method is based on maxterms but, as two different definitions of a maxterm are in common use, there are two different maxterm representations of a circuit.

Definition 1 A **maxterm** is an OR function which includes every input variable (literal) once only in either true or complemented form. The maxterm can only have the value 0 when all the variables have values which are the inverse of those in the row of the truth table described by the maxterm.

Definition 2 A **maxterm** is that OR function which contains every literal once only in either true or complemented form. The OR function has the value 0 only if all the literals have the same values as those in the row of the truth table described by the maxterm. Table B.1 lists all the possible values of three variables A, B and C with the corresponding maxterms according to both definitions.

Table B.1

C	B	A	Definition 1	Definition 2
0	0	0	$\bar{A}+\bar{B}+\bar{C}$	$A+B+C$
0	0	1	$A+\bar{B}+\bar{C}$	$\bar{A}+B+C$
0	1	0	$\bar{A}+B+\bar{C}$	$A+\bar{B}+C$
0	1	1	$A+B+\bar{C}$	$\bar{A}+\bar{B}+C$
1	0	0	$\bar{A}+\bar{B}+C$	$A+B+\bar{C}$
1	0	1	$A+\bar{B}+C$	$\bar{A}+B+\bar{C}$
1	1	0	$\bar{A}+B+C$	$A+\bar{B}+\bar{C}$
1	1	1	$A+B+C$	$\bar{A}+\bar{B}+\bar{C}$

A detailed comparison of the two definitions is only relevant to those concerned with the development of methods for logic circuit design. Engineers applying such design methods will usually find that the techniques already described which use minterms are the most useful ones. There is a marginal advantage in using a maxterm approach when circuits constructed from NOR gates are required rather than circuits using NAND gates.

The two different maxterm definitions lead to two different methods of circuit design.

Definition 1 is useful when generalized theorems of Boolean algebra are adopted but Definition 2 is the more simple one to apply to the design of combinational logic circuits. A brief outline showing the development of a minimal expression for a logic circuit using maxterms based on Definition 2 follows.

Table B.2 is the truth table of a three-input circuit which is described in minterms by

$$R = \bar{A}.\bar{B}.\bar{C} + \bar{A}.B.\bar{C} + \bar{A}.B.C$$

This is easily reduced to the sum of products form

$$R = \bar{A}.B + \bar{A}.\bar{C}$$

Table B.2

Inputs			Output	Maxterm	Minterm
C	B	A	R	(Definition 2)	
0	0	0	1	$A+B+C$	$\bar{A}.\bar{B}.\bar{C}$
0	0	1	0	$\bar{A}+B+C$	$A.\bar{B}.\bar{C}$
0	1	0	1	$A+\bar{B}+C$	$\bar{A}.B.C$
0	1	1	0	$\bar{A}+\bar{B}+C$	$A.B.\bar{C}$
1	0	0	0	$A+B+\bar{C}$	$\bar{A}.\bar{B}.C$
1	0	1	0	$\bar{A}+B+\bar{C}$	$A.\bar{B}.C$
1	1	0	1	$A+\bar{B}+\bar{C}$	$\bar{A}.B.C$
1	1	1	0	$\bar{A}+\bar{B}+\bar{C}$	$A.B.C$

Using Definition 2 the maxterms for which the circuit output R is 0 are $\bar{A}+B+C$, $\bar{A}+\bar{B}+C$, $A+B+\bar{C}$, $\bar{A}+B+\bar{C}$ and $\bar{A}+\bar{B}+\bar{C}$. Each maxterm in this list can only be 0 for input conditions corresponding to the values of the inputs in the row of the truth table identified by the maxterm. As a 0 dominates an AND function and R is 1 unless one of the listed maxterms is 0 then R is the AND function of all these maxterms, i.e.

$$R = (\bar{A}+B+C).(\bar{A}+\bar{B}+C).(A+B+\bar{C}).(\bar{A}+B+\bar{C}).(\bar{A}+\bar{B}+\bar{C})$$

A Karnaugh map with the squares identified by maxterms is shown for 3-variables in Fig. B.1a and the completed map for the example is Fig. B.1b. As the method is a dual of the minterm technique groups of zeros are formed on the map instead of groups of ones.

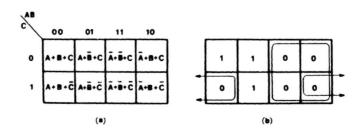

(a)   (b)

Fig. B.1

The variables which change within a group are eliminated and the groups are described by sum terms which are then combined in a product expression. In Fig. B.1b the group of four is simply $\bar{A}$ and the group of two is $B + \bar{C}$; these combine giving

$$R = \bar{A}.(B + \bar{C})$$

This result is a product of sums expression and it is easily converted into the sum of products result obtained using minterms. Alternatively, double inversion and application of de Morgan's theorem leads to the NOR function form

$$R = \overline{A + (\overline{B + \bar{C}})}$$

Appendix C

A Note Concerning Symbols

By using the older, distinctively shaped symbols as 'pure logic' symbols, many of the problems of representing complete integrated circuit logic devices by symbols have been ignored. The 'new' symbols of the International Electrotechnical Commission (IEC) in publication 617 Part 12 are used by most national standards bodies to form their own specifications. Generally these are identical to the IEC publication with the addition of explanatory material. In the UK, the standard is BS3939: Part 12 (1985). In the USA it is ANSI/IEEE Std 91–1984.

As these standards are intended to represent complete electrical logic circuits in an exact manner they are complex. Also problems arise in understanding symbols because any system adopted must not lead to confusion between the signal (usually a voltage) used to represent a logic level, and the name given to the logic level. It was indicated in Chapter 2 that a decision must be made as to which of the two voltage levels used in any practical logic system is called 0 and which is called 1. The common practice is to use a positive logic convention with the upper voltage level representing 1 and the lower level representing 0. The converse, negative logic convention is equally valid. A 'pure logic' circuit will function according to the logic definition regardless of convention, the actual level for each signal is only required when the measurement of the level is made to determine the logic level.

Once electrical measurements have to be related to logic values a complex system of symbols is required to produce the exact definition. The NAND gate symbols of Chapter 2 are reproduced here as Fig. C.1. The small circle used in parts (a), (b) and (c) indicates a logic inversion, whereas the small triangle used in part (d) is a low polarity indicator. On an initial examination the triangle appears to perform the same function as the circle but its purpose is to indicate that the active voltage level is low.

Part (d) of Fig. C.1 is more precise than the other parts of the Figure. The box by itself indicates an AND function; that is if the inputs are all active (at the higher of the two voltages used in the logic system) then the output of the box will be active (at the higher voltage level). However, the box has a low polarity indicator at its output in Fig.

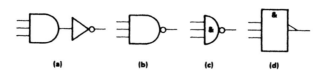

(a)   (b)   (c)   (d)

Fig. C.1 NAND gate symbols

C.1d, this means that the output is active when at the low voltage level, not at the high level. Thus the symbol in Fig. C.1d is not strictly that for a logic NAND gate; it actually implies that if all the inputs are at the higher of the two voltage levels then the output will be at the lower level and in all other cases the output will be at the higher level. If we choose to call the upper level logic 1 the device functions as a NAND gate. (Note that low polarity indicators may be used at inputs to elements as well as at outputs.)

The full standard is a long and complicated document; however the complexity is necessary in order to define symbols for all possible situations and for precisely specifying complete devices. The reader is advised to become familiar with the basic concepts of logic circuit design before attempting to understand the IEC symbols. When the symbols are used the relevant standard for the country in which the reader is working should be used. The bibliography lists one book which further explains the symbols. It is also useful to examine manufacturers' data books for examples of use of the symbols, as these books will also show how 'pure logic' symbols and the standard device symbols are used in different ways.

Appendix D

Solutions to Selected Problems

If a problem has a single definite answer only the final result is given. If cases for which more than one solution is possible a brief outline of a solution is provided; this particularly applies to problems which require the proof of a relationship or the design of a circuit.

Chapter 1

2 (a) 101100101; 1000100; 111011·10111000 (to 8 places); 10111·011
(b) 26; 11; 119·625; 42·375 (c) 190 (d) 356
(e) 1111001010; 11110100·01111101 (f) 3FD; 3B·B8 (to 2 places)

4 Three bits (octal digit). Four bits (duodecimal digit). A simple weighted code is commonly used and is shown in the following table.

Digit	Usual symbol	Binary code
zero	0	0000
one	1	0001
two	2	0010
three	3	0011
four	4	0100
five	5	0101
six	6	0110
seven	7	0111
eight	8	1000
nine	9	1001
ten	A	1010
eleven	B	1011

Chapter 2

1 (a) 1 (c) 0

3 (a) $\bar{C}$ (b) $\bar{L}.M + M.\bar{N}.P + \bar{L}.\bar{N}.\bar{P}$ (c) $\bar{B}.D + B.\bar{C}.\bar{D} + \bar{A}.C.\bar{D}$

5 (a) Use the definition of the exclusive-OR function and develop tables to show the required identities.

A	B	A⊕B	Ā	B̄	Ā⊕B̄
0	0	0	1	1	0
0	1	1	1	0	1
1	0	1	0	1	1
1	1	0	0	0	0

Clearly the columns for $A \oplus B$ and $\bar{A} \oplus \bar{B}$ are identical for all possible values of A and B.
  (b) Similarly

A	B	A⊕B	$\overline{A \oplus B}$	Ā	B	$\bar{A} \oplus B$	A	B̄	$A \oplus \bar{B}$
0	0	0	1	1	0	1	0	1	1
0	1	1	0	1	1	0	0	0	0
1	0	1	0	0	0	0	1	1	0
1	1	0	1	0	1	1	1	0	1

Again the columns for the required quantities are identical.

Chapter 3

2 Assume that a detector gives an output of 1 if a fault is present. Choose an alarm to be indicated when the circuit output is 1.

Inputs	Output
0000	0
0001	0
0010	0
0011	1
0100	0
0101	1
0110	1
0111	1

Inputs	Output
1000	0
1001	1
1010	1
1011	1
1100	1
1101	1
1110	1
1111	1

4 The circuit equations are

$$X = A_1 . \bar{B}_1 + A_0 . A_1 . \bar{B}_0 + A_0 . \bar{B}_0 . \bar{B}_1$$
$$Z = \bar{A}_1 . B_1 + \bar{A}_0 . \bar{A}_1 . B_0 + \bar{A}_0 . B_0 . B_1$$
$$Y = \bar{A}_0 . \bar{A}_1 . \bar{B}_0 . \bar{A}_1 + A_0 . \bar{A}_1 . B_0 . \bar{B}_1 + \bar{A}_0 . A_1 . \bar{B}_0 . B_1 + A_0 . A_1 . B_0 . B_1$$
$$= \bar{X} . \bar{Z}$$

6 Solutions depend upon several choices to be made by the designer; for example is digit 6 displayed with or without segment 'a' (the top horizontal bar)? Taking segments 'b' and 'e' as examples with A, B, C and D the binary inputs (A as least significant digit) and unused codes taken as don't care conditions then:

$$b = A.B + \bar{A}.\bar{B} + \bar{C}$$
$$e = \bar{A}.B + \bar{A}.\bar{C}$$

Chapter 4

2 One solution is shown in outline in Fig. D.1. If the direction control is 0 for upward shifting and 1 for donward shifting then the complete solution for the combinational circuit is described by the equation

$$I_N = \bar{K}.Q_{N-1} + K.Q_{N+1} = (\overline{\overline{K}.Q_{N-1}}).(\overline{K.Q_{N-1}})$$

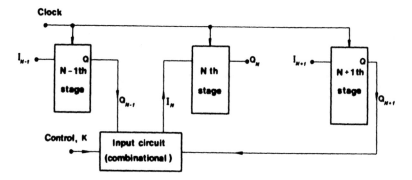

Fig. D.1

5 Tabulated results

Clock pulses	A	B	C	D	Next input = $\bar{D}$
Initial state	0	0	0	0	1
1	1	0	0	0	1
2	1	1	0	0	1
3	1	1	1	0	1
4	1	1	1	1	0
5	0	1	1	1	0
6	0	0	1	1	0
7	0	0	0	1	0
8	0	0	0	0	1
9 (= 1)	1	0	0	0	1
10 (= 2)	1	1	0	0	1

This is four stage Johnson counter (see Section 8.4).

Chapter 5

2 An extended form of Fig. 5.3 shows that counter states 000 and 100 exist for less time than any others. If the clock input has period T and all delays are t_d then these two states exist for a time $T - 2 \times t_d$. Substituting values requires $T - 2 \times 20 \geqslant 20$ (units are nanoseconds) giving T≥60 nsec for useful operation of the counter. Therefore the counter is limited to use at frequencies below 16·7 MHz.

Chapter 6

3 The state diagram (not shown) is just a simple ring of the states. The question requires that the state allocation has S_1 as $A = B = C = 0$; S_2 as $A = 1$, $B = C = 0$; S_3 as $A = B = 1$, $C = 0$; S_4 as $A = C = 1$, $B = 0$; and S_5 as $A = 0$, $B = C = 1$.

Taking unused states as having 'don't care' control conditions the solution for the control inputs is $J_A = \bar{B}$, $K_A = C$, $J_B = A$, $K_B = 1$, $J_C = K_C = B$.

(There is an alternative solution for J_A of $J_A = C$ but it can stick at switch on.) The circuit diagram is shown in Fig. D.2.

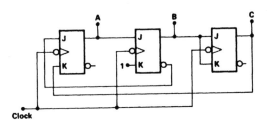

Fig. D.2

7 The design procedure is exactly the same as Example 6.4 but the excitation table for an SR flip-flop must be used instead of that for a JK device.

Present output Q_n	Next output Q_{n+1}	Control required	
		S	R
0	0	0	X
0	1	1	0
1	0	0	1
1	1	X	0

Figure D.3 is the circuit diagram of the divide-by-four up-down counter.

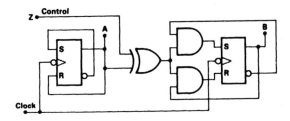

Fig. D.3

Chapter 7

2 For cases as simple as this 'transient free' circuits can be designed but the problem specifies that a solution is required using transient elimination techniques (often used with large networks). One solution is illustrated in Fig. D.4; this has input and output latches with the output latch loaded at a time when no transients can exist.

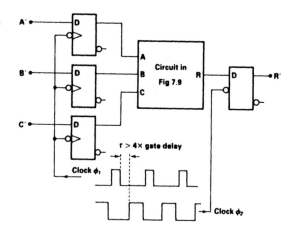

Fig. D.4

4 The input signals are connected to individual buffers whose outputs are connected in parallel to the transmission line. As only one buffer may have its tri-state output active at any time the enable inputs must be controlled by a circuit which ensures this. This control circuit is '1 out of 4 decoder' similar to that used in Section 5.4; the complete circuit is shown in Fig. D.5 (it is assumed that the tri-state outputs are active when the enable input is at 1, not when 0 as in Fig. 7.4).

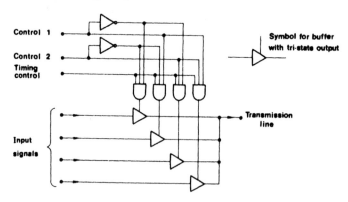

Fig. D.5

Chapter 8

1   There are several solutions, one is to copy the ripple-carry adder but use a different cell. This will have two digits, A_i and B_i, and a borrow-in C_{i-1} as inputs; it will produce a difference D_i and a borrow-out C_i. The cell truth table for $A - B$ is:

B_i	A_i	C_{i-1}	D_i	C_i
0	0	0	0	0
0	0	1	1	1
0	1	0	1	0
0	1	1	0	0
1	0	0	1	1
1	0	1	0	1
1	1	0	0	0
1	1	1	1	1

If the full-adder cells of Fig. 8.4 are replaced by cells designed to satisfy this truth table then the complete circuit will subtract number B from number A.

2   Hint. Prepare a truth table and draw a Karnaugh map for a parity indicating circuit for a 3-bit number. Repeat for a 4-bit number. These maps should suggest that there is a simple progression from an n-bit circuit to an $(n+1)$-bit circuit and the cell required should be obvious.

Bibliography

The following list is a selection of books which are concerned with logic circuits and digital computers; most of the books are more advanced than this one. Several of the books at the end of the list are published by integrated circuit manufacturers and usually they can only be obtained from the manufacturers or their appointed distributors.

Bannister, B. R. and Whitehead, D. G. *Fundamentals of Modern Digital Systems*. (2nd Edition). Macmillan 1987

Bolton, M. *Digital System Design with Programmable Logic*. Addison-Wesley 1990

Bostock, G. *Programmable Logic Handbook*. Collins 1987

Clements, A. *The Principles of Computer Hardware*. Oxford University Press 1985

Comer, D. J. *Digital Logic and State Machine Design*. Holt Reinhart and Winston 1984

Floyd, T. L. *Digital Fundamentals*. (2nd Edition). Merrill 1982

Gibson, J. R. *Electronic Processor Systems*. Edward Arnold 1987

Gosling, J. B. *Digital Timing Circuits*. Edward Arnold 1985

Hill, F. J. and Peterson, G. R. *Digital Logic and Microprocessors*. Wiley and Sons 1987

Kampel, I. J. *A Practical Introduction to the New Logic Symbols*. Butterworths 1986

Lewin, D. *Design of Logic Systems*. Van Nostrand Reinhold 1985

Nashelsky, L. *Introduction to Digital Technology*. (3rd Edition). Wiley and Sons 1983

Philips Components. *Technical Handbook*, Book 4 Part 5, *Integrated Circuits*. Philips Components Ltd 1988

Texas Instruments. *The TTL Data Book*, Volume 2. Texas Instruments Inc 1989

Index

Absorption rules 21
Access time 117
Action table 58, 79
Active high 101
Active low 101–102
Addend 107–111
Addition 107–111
Address 117–118
AND **12–14**, 18–24
Associative law 21
Asynchronous circuits 48, 66–72, 112–113
Asynchronous counter 66–72, 112
Augend 107–111

Base 5, 6, 129–132
Binary coded decimal (BCD) 7, 42
Binary counter 67
Binary number system 6, 129–132
Bistables **48**, 48–63
Bit 6, 53, 117
Boole G. 18
Boolean algebra 14–18, 20–26
Boolean arithmetic 18–20
Boolean variable 20
Buffer 97
Bus 100

Carry 107–110
Cell technique 111–116, 124–125
CLEAR 60
Clock input 52–63, 75
Ciock pulse **52**, 52–62, 66, 68
Clocked SR flip–flop 51–53
CMOS 4, 94–100
Codes 6–8
Commutative laws 21
Complement 15, 21
Contact bounce 103
Continuous-state system 1
Control input 57, **75**, 75–86
Counters **66–73**, 83–91, 112–116

Decimal number system 5–6, 129–131
de Morgan's theorem 21–23, **23–40**, 135
Digit **5–6**, 108
Discrete-state system 1
Distributive laws 21
Divide-by-N counter 66–71, 112–116

Don't care **41–44**, 79–81
Down counter 68
D-type flip-flop 53–54
Dual in line (DIL) package 4
Duality 23–24, 133–134
Duodecimal number system 9, 131

Edge-triggered 53, 60–64
Electronic logic components 2–5, 94
Emitter coupled logic (ECL) 4
Enable input 101
Excess three code 8
Excitation table 78–89
Exclusive-NOR 27
Exclusive-OR 24–26, 44, 139

FALSE 3, 18
Fan-in 97–100
Fan-out 97–100
Feedback 48
Flip-flop **48–53**, 57–64
Full-adder 108–111

Gate 12, 12–18
Gray code 7–8, 68

Half-adder 108–110
Hazard **50**, 52, 62, 71, 94–97
Hexadecimal number system 8, 132
Hold time 61–62

Idempotence 21
Input conditions 75–78
Integrated circuit 3–5, 117, 120
Interface circuits 4, 100, 102
INVERTER **15**

JK flip-flop **57–64**, 69, 77–80
Johnson counter 114–115

Karnaugh map 34–44, 59, 82–87, 134

Latch 53
Light emitting diode (LED) 104
Literal 20, 31, 133
Logic circuit analysis 26–27
Logic elements 2, 12–18
Logic operators **19–20**

Logic networks 2
Logic systems 2

Master-slave 55–56, 58–60
Maxterm **32**, 133–135
Memory 48, 117–118
Minimization 33–42, 127
Minterm **31–33**, 133–134
Moore's model 74
Multiple-output circuits 10–11, 44–45

NAND **16–17**, 23, 39–41
Non-binary counters 67
NOR **17–18**, 39–41
NOT **15–19**
Numbers 5–6, 129–132

Octal number system 8, 131–133
ONE **3**, 6, 18–21
OR **14–15**, 18–20
Oscillator 95

Parallel addition 110–114
Parallel in parallel out (PIPO) register 57, 73, 116
Power consumption 4, 99
Precedence 19–20
PRESET 60, 67
Product term 32
Product of sums 135
Program 66
Programmable logic device (PLD) 121–128
Programmable read only memory (PROM) 117, 122
Propagation delay 10, 48, 62, 70–72, 94–99
Pulse **52**, 68
Pulse sequences 68
Pure binary counters 67–70

Race condition **50**, 52, 61–63, 94–95
Radix 5
Random access memory (RAM) 117
Read only memory (ROM) 117, 122
Registers 54–57
Reset 49, 52, 58, 67
Ripple counter 70

Serial addition 110–111
Serial in-parallel out (SIPO) register 56
Serial in-serial out (SISO) register 56
Serial shift register 54–57, 114–15

SET 49, 52, 58
Set-reset (SR) flip-flop 48–53, 95
Set-up time 61
Seven segment indicator 46–47
State assignment 74–78
State diagram 75–77
State of a system 1, 1–3, 6, 49, 74–78
State table 77–78
Storage element 53, 117
$\overline{\text{SR}}$ flip-flop 50–51, 103
Sum of products 33, 39, 44
Sum term 32
Switching circuits 2
Switching table 52
Synchronous circuits 48, 52
Symbols 13–18, 25, 51, 136–137
System 1

Timing diagram 69–70, 90–92
Toggle action 58, 64
Transient 71–73, 95–97
Transient elimination 72–73, 90
Transistor-transistor logic (TTL) 3, 4, 94, 98
Transition 79
Transition table 52, 77
Tri-state 101–102
TRUE 3, 18
Truth table **11**, 29–31
T-type flip-flop 64
Twisted ring counter 114
Two-phase clock **55**, 72, 97
Two-state systems 1–3, 6

Unit 6, 21
Unit load 97
Unit rule 21
Unspecified states 41–43
Unused states 80, 87–89
Up counter 68
Up-down counter 68

Variable 21–23, 31, 35, 133
Venn diagram 34

Weighted code 6
Wired-OR 101
Word 54, 117

ZERO **3**, 6, 18–21
Zero rule 21